T0192396

Critical Animal Geographies

Critical Animal Geographies provides new geographical perspectives on critical animal studies, exploring the spatial, political, and ethical dimensions of animals' lived experience and human–animal encounters. It works toward a more radical politics and theory directed at the shifting boundary between human and animal. Chapters draw together feminist, political-economic, posthumanist, anarchist, postcolonial, and critical race literatures with original case studies in order to see how efforts by some humans to control and order life – human and not – violate, constrain, and impinge upon others. Central to all chapters is a commitment to grappling with the stakes – violence, death, life, autonomy – of human–animal encounters. Equally, the work in the collection addresses head-on the dominant forces shaping and dependent on these encounters: capitalism, racism, colonialism, and so on. In doing so, the book pushes readers to confront how human–animal relations are mixed up with overlapping axes of power and exploitation, including gender, race, class, and species.

Kathryn Gillespie is a Postdoctoral Fellow in Animal Studies at Wesleyan University, USA. Her research focuses on the lived experience of animals in spaces of commodity production (e.g., farming, breeding, sale, and slaughter), with a particular emphasis on those animals humans use for food.

Rosemary-Claire Collard is an Assistant Professor in Geography at Concordia University in Montreal, Canada. Her research looks at capitalism, environmental politics, science, and culture, especially film, with an eye to how they depend on and engender certain human–animal relations.

Routledge Human–Animal Studies Series

Series edited by Henry Buller

Professor of Geography, University of Exeter, UK

The new *Routledge Human–Animal Studies Series* offers a much-needed forum for original, innovative and cutting edge research and analysis to explore human–animal relations across the social sciences and humanities. Titles within the series are empirically and/or theoretically informed and explore a range of dynamic, captivating, and highly relevant topics, drawing across the humanities and social sciences in an avowedly interdisciplinary perspective. This series will encourage new theoretical perspectives and highlight groundbreaking research that reflects the dynamism and vibrancy of current animal studies. The series is aimed at upper-level undergraduates, researchers, and research students as well as academics and policymakers across a wide range of social science and humanities disciplines.

Published

Critical Animal Geographies
Politics, intersections, and hierarchies in a multispecies world
Edited by Kathryn Gillespie and Rosemary-Claire Collard

Forthcoming

Urban Animals
Crowding in zoocities
Tora Holmberg

Critical Animal Geographies

Politics, intersections, and hierarchies in a multispecies world

Edited by
Kathryn Gillespie and
Rosemary-Claire Collard

Routledge
Taylor & Francis Group

LONDON AND NEW YORK

First published in paperback 2017

First published 2015
by Routledge
2 Park Square, Milton Park, Abingdon, Oxon, OX14 4RN

and by Routledge
711 Third Avenue, New York, NY 10017

Routledge is an imprint of the Taylor & Francis Group, an informa business

British Library Cataloguing in Publication Data
A catalogue record for this book is available from the British Library

Library of Congress Cataloging in Publication Data
Collard, Rosemary-Claire.
Critical animal geographies / Rosemary-Claire Collard, Kathryn Gillespie.
   pages cm. – (Routledge human-animal studies series)
   1. Animals and civilization. 2. Animals–Effect of human beings on.
   I. Gillespie, Kathryn (Kathryn A.) II. Title.
   QL85.C645 2015
   591.5–dc23   2014031974

ISBN: 978-1-138-79150-3 (hbk)
ISBN: 978-1-138-63470-1 (pbk)
ISBN: 978-1-315-76276-0 (ebk)

Typeset in Times New Roman
by Wearset Ltd, Boldon, Tyne and Wear

Contents

Figures

Contributors

Shelley Alexander is an Associate Professor in the Department of Geography at the University of Calgary, Canada. She has conducted field research and GIS modeling for carnivores in the Canadian Rockies since 1990. With an emphasis on wolves and coyotes, her research spans canid ecology, human–wildlife conflict, road ecology, media content, and spatial analysis. Dr. Alexander established the Calgary Coyote Project (2005), examining ecology and human dimensions of human–coyote interactions, and spearheaded Living with Coyotes (an online citizen science tool). Other research collaborations include modeling swift fox critical habitat, road effects on large carnivores (Yucatan, MX), and Painted Dog conservation (Zimbabwe).

Gwendolyn Blue is an Assistant Professor in the Department of Geography at the University of Calgary, Canada. Formally trained in cultural studies, her research centers on manifestations of the public in posthumanist contexts in which the boundaries between humans, animals, and technologies have been called into question. This inquiry takes shape in two domains: public participation in environmental governance, with a focus on climate change, and public engagement with food politics, with a focus on the politics of flesh consumption. Publication topics range from food safety concerns (BSE), human–animal relations, to formal public engagement with climate change.

Irus Braverman is Professor of Law and Adjunct Professor of Geography at SUNY Buffalo. Her main interests lie in the interdisciplinary study of law, geography, and anthropology. Braverman published *House Demolitions in East Jerusalem: 'Illegality' and Resistance* (Hebrew, Steinmetz), *Planted Flags: Trees, Land, and Law in Israel/Palestine* (Cambridge University Press, 2009), *Zooland: The Institution of Captivity* (Stanford University Press, 2012), *The Expanding Spaces of Law: A Timely Legal Geography* (co-edited with Nicolas Blomley, David Delaney, and Alexandre Kedar, Stanford University Press, 2014), and *Wild Life: The Institution of Nature* (Stanford University Press, 2015).

Rosemary-Claire Collard is an Assistant Professor in Geography at Concordia University in Montreal. Her research looks at capitalism, environmental politics, science, and culture, especially film, with an eye to how they depend on and engender certain human–animal relations. Her most recent research project tracked the exotic pet trade through six countries and is forthcoming as a book, *Animal Traffic* (Duke University Press).

Jody Emel is a Professor of Geography at Clark University in Worcester, Massachusetts. She is co-editor with Jennifer Wolch of *Animal Geographies: Place, Politics and Identity in the Nature-Culture Borderlands*. A long-time animal rights activist, she is author of multiple publications critical of the industrial meat production/consumption regime. She is currently writing a book with Harvey Neo entitled *Geographies of Meat: Culture, Politics, Economy* (Ashgate) and editing another (also with Harvey Neo) on the political ecologies of meat for Earthscan.

Kathryn Gillespie is a part-time Lecturer in Geography, the Honors Program, and the Comparative History of Ideas Program at the University of Washington in Seattle, Washington. Her research focuses on the lived experience of animals in spaces of commodity production (e.g., farming, breeding, sale, and slaughter), with a particular emphasis on those animals humans use for food. She is currently working on a book, *The Cow with Ear Tag #1389*, based on her most recent research on the lives of cows in the U.S. dairy industry.

Eva Giraud is a Lecturer in Media, Communication and Culture at Keele University. Her research interests focus on posthumanism; critical animal geographies; new social movement studies and technologically mediated activism. She is particularly interested in the political value of posthumanism, and – conversely – what theory can learn from contemporary grassroots activism. Research developing these themes has been published in journals including *Culture, Theory and Critique*, *Convergence*, *PhaenEx*, and *Radical Philosophy*, and she is currently working on a monograph about posthumanist politics.

Connie L. Johnston is a Visiting Instructor at the University of Oregon in Eugene. Connie is interested in the myriad ways in which human relations with nonhumans are mediated by scientific knowledge, and she pursued this area of interest as a research fellow in the Harvard Science, Technology & Society Program in 2012–2013. Her dissertation supported by a National Science Foundation grant, examined the role of government-supported farm animal welfare science in the U.S. and Europe. A portion of this research is presented in her article 'Geography, Science, and Subjectivity' (*Geography Compass*, 2013) and a chapter in the forthcoming book *The Political Ecology of Meat* (J. Emel and H. Neo, eds.). Her current work focuses on furthering posthumanist political theory.

John Joyce is currently a Master in Teaching student at The Evergreen State College in Olympia, Washington. He previously taught middle school Science and Social Studies in Browning, Montana on the Blackfeet Indian Reservation. John graduated from Vassar College in 2012 with a BA in Latin

American and Latino/a Studies and a correlate sequence in Geography. He was an active member of the Vassar Animal Rights Coalition and the Vassar Association of Class Activists. His senior thesis focused on the dynamic relationship between migrant labor, ducks, and commodities in relation to Hudson Valley Foie Gras.

Karen M. Morin is Associate Dean of Faculty and Professor of Geography at Bucknell University in Lewisburg, Pennsylvania, USA. Her interests span the history of geographical thought and literacy in North America, nineteenth-century travel writing, postcolonial geographies, the geography of religion, and most recently, critical prison studies. Her most recent book is *Civic Discipline: Geography in America, 1860–1890* (Ashgate, 2011). She is a decade-long executive board member of a local nonprofit prisoner rights group, the Lewisburg Prison Project.

Heidi J. Nast is Professor of International Studies at DePaul University, Chicago. Her work examines the political economy of fertility and reproduction in relation to critical theories of sexuality, race, and gender as well as critical social, economic, and cultural theory. She has published widely across disciplines and is currently working on a three-volume work, *Petifilia*, the first volume of which is forthcoming with University of Georgia Press.

Joseph Nevins is an Associate Professor in the Department of Earth Science and Geography at Vassar College in Poughkeepsie, New York. His research interests include socio-territorial boundaries and mobility, violence and inequality, and political ecology. Among his books are *Operation Gatekeeper and Beyond: The War On 'Illegals' and the Remaking of the U.S.–Mexico Boundary* (Routledge 2010); *Taking Southeast Asia to Market: Commodities, Nature, and People in the Neoliberal Age* (co-edited with Nancy Peluso, Cornell University Press, 2008); and *A Not-so-distant Horror: Mass Violence in East Timor* (Cornell University Press, 2005).

Claire Rasmussen is Associate Professor of Political Science and International Relations. Her book *The Autonomous Animal: Self-Governance and the Modern Subject* appeared in University of Minnesota Press in 2011. Her work, which includes examinations of subjectivity, sexuality, and animality has appeared in places including *Signs: A Journal of Women and Culture, Society and Space, Social and Cultural Geography*, and *Citizenship Studies*. She is currently finishing a manuscript on the relationship between liberal citizenship and the regulation of sexuality and is working on the idea of the political nature of animals.

Jill S. Schneiderman is a Professor in the Department of Earth Science and Geography at Vassar College where she teaches transdisciplinary courses in the programs in Women's Studies; Science, Technology and Society; and Environmental Studies. She is editor of and contributor to three books: *Liberation Science: Putting Science to Work for Environmental Justice* (2012); *For the Rock Record: Geologists on Intelligent Design* (2009); and *The Earth*

Around Us: Maintaining a Livable Planet (2003), as well as the author of numerous scientific papers concerned with geologic changes reflected in recent sediments of the Yangtze and Nile deltas. She was a Fulbright Fellow at the University of the West Indies at the Centre for Gender and Developments Studies, Trinidad and Tobago.

Elisabeth (Lisa) Stoddard is an Assistant Teaching Professor in Undergraduate Studies and Social Science and Policy Studies at Worcester Polytechnic Institute in Massachusetts. Her research focuses on the policy and politics of food production in a changing climate and global economy. She analyzes the ways in which the governance of agriculture and food animal production shapes our food systems' and communities' vulnerability and capacity to adapt to drought, floods, the global spread of disease, and other hazards. She is also interested in issues of human–nonhuman injustice in the livestock industry and the ability of social movements to make powerful changes. Stoddard is a Switzer Environmental Fellow, a member of the North Carolina Environmental Justice Network, and she collaborates with the Waterkeeper Alliance's Concentrated Animal Feeding Operation Division in North Carolina.

Richard J. White is a Senior Lecturer in Human Geography based at Sheffield Hallam University, UK. A significant part of his teaching, research, and writing is committed to developing anarchist praxis within human geography. Addressing a range of ethical landscapes rooted in the context of social justice and total liberation movements, he is particularly interested in deconstructing the ways in which exploitation of humans and animals runs side-by-side and intersects in society, and developing a new geographic imaginary based on peace and nonviolence. He has recently published chapters in *The Accumulation of Freedom: Writings on Anarchist Economics* (AK Press) and *Defining Critical Animal Studies: an Intersectional Social Justice Approach for Liberation* (Peter Lang Publishing Group). He was the editor of the *Journal for Critical Animal Studies* between 2009 and 2012.

Anastasia Yarbrough is a facilitator and consultant in the areas of community building, and diversity and inclusion. She has been an activist for animal liberation and ecological justice since 2003. She holds a bachelor's degree in Natural Resources with a focus in wildlife ecology, and wildlife and society. She is a former board member for the Institute for Critical Animal Studies and former editor for a special issue of the *Journal for Critical Animal Studies*. She volunteers regularly with neighborhood associations to design and implement wildlife conflict mediation strategies that encourage bioculturally diverse communities and are life-affirming.

Acknowledgments

We would like to thank Faye Leerink for her keen editorial role in producing this book, and Henry Buller as the editor of the *Human–Animal Studies Series* at Routledge. We are also grateful to Joanna Reid and three anonymous reviewers for their invaluable assistance at the proposal stage. We thank the authors who have labored to research, write, and revise their thoughtful and provocative chapters for this volume. It has been a pleasure to engage with such smart and dedicated people. And finally, we acknowledge and grieve the animals who are currently living, laboring, held captive, and dying in service to capital accumulation and human dominance.

1 Introduction

Rosemary-Claire Collard and Kathryn Gillespie

The cow lies on the manure-covered floor of a holding pen at the dairy market sale behind a farmed animal auction yard in California's Central Valley. Her past as a typical dairy producer is easily visible on her body in the scars on her hide, the blood and milk leaking from her udders, her docked tail and ear tags, and her emaciated appearance. A sticker with a barcode is stuck haphazardly to her side with #743, her identifying numbers at the auction. She lies on her stomach, her back legs splayed out behind her, unable to rise. She struggles several times to get to her feet and each time collapses with exhaustion, breathing heavily, saliva foaming at her mouth. Other cows mill around her until an auction employee comes and herds the others into the adjacent pen. He binds her rear ankles together with a rope and prods her to stand. She is unable to rise. The employee leaves her in the pen. She will not be sold at the auction. Because she cannot stand, she will be categorized as a 'downer' and shot with a firearm at the end of the day. A 'deadstock' hauler will be called to pick up her body and deliver her to a rendering facility where her body will be reduced to useable commodity products.

...

In the packed stands of a hot barn, hundreds of auction goers at the Lolli Bros Livestock Market Exotic Sale in Macon, Missouri, lean forward in their seats to glimpse the next round of objects wheeled out into the auction ring for sale. It is a line-up of beige plastic cages. People quiet down until all you can hear are the muffled squawks and shrieks of hundreds of parrots and other birds in an adjacent room, having already been sold. The cages now in the ring look like medium-sized dog kennels, stacked two on two. A buzz of excitement ripples through the audience when long, thin, dark fingers slip out through the slats in several cages. Half a dozen adult spider monkeys are crouched in cages too cramped for them to stand. The first monkey on the auction block is a nineteen-year-old female who lost her right eye eight years ago and had two middle fingers half amputated. "But all her lady parts work," the auctioneer assures the crowd. She sells for $2,500. These monkeys are all 'proven breeders,' no longer small and docile enough for display, or to be cuddled, diapered and fitted with pink dresses and camouflage overalls. Their value is as 'breed-stock,' producing

$10,000 babies to be sold at next year's auction, maybe the one in Texas, or Ohio.

Taken from our research on political economies of cows in the dairy industry (KG) and exotic pets (RC), these two snapshots introduce the project of this book: to understand the spatial, political-economic, and ethical dimensions of animals' lived experience and human–animal encounters. In the spirit of this project, these vignettes focus on animals as they move through the live animal auction – a power-laden space that determines animals' fate at the point of sale. They also draw attention to the violent power relations at work in human–animal relations. This violence is carved into the bodies of the cow with barcode #743 and the captive spider monkey, visible in their physical degradation and the cow's inability to rise from the floor of the auction yard. This violence is standard, woven into day-to-day operations in these facilities and economies. It goes unremarked upon, let alone is it mourned.

Live animal auctions in the United States are places where the geographical dimensions of power and hierarchy between humans and other animals can be seen in stark detail. Nonhuman animals are subjected to various modes of bodily control in the space of the auction yard where they are exchanged as commodities and used in the production of new commodities. Animals sold at auction come from myriad conditions, move through the auction space in strategically manipulated states of (im)mobility and captivity, and are destined for a range of uses as they leave the auction yard. Farmed animals, like the cow with barcode #743, are sold and bought at auction to be used as commodity producers (e.g., for breeding, milk production, semen production) and as commodities themselves (e.g., to be slaughtered for 'meat'). Even animals raised on small-scale so-called humane farms are often subjected to the violence of the auction yard. Animals at the exotic animal auctions, like the nineteen-year-old spider monkey, are sold and purchased as companions; as spectacles in petting zoos, menageries or private game ranches; as breeders; as performers in the film and entertainment industry; and as scientific testing subjects, with some of these uses overlapping at any one time and several animals passing through several of these roles over the course of their lifetimes.

At farmed and exotic animal auctions, animals are transported – sometimes from across the United States – and collected in one space for a day or two (or even just a few hours) before being re-dispersed around the country. Farmed animals arrive at auction in transport trailers and are unloaded and herded into a series of chutes and holding pens behind the auction yard. Here, they wait while potential buyers walk through and assess the value of their bodies and (re)productive potential. Sometimes farmed animal auctions are single-species auctions, organized by their commodity potential (e.g., dairy market, cull market, feeder sales, etc.); other times they are multispecies auctions, attracting a more diverse crowd with interest in various farmed animal species. Exotic pet auctions are typically held in farmed animal auction yards. Smaller animals, like monkeys or snakes, are sold in cages, whereas larger animals, like zebras or camels, are herded through the auction ring like cows at the farmed animal auction. Where farmed animal auctions are mundane, ordinary places, the exotic pet auction transforms these venues into places of spectacle and display. Audiences of thousands gather to pace the catwalk above a sea of penned zebras and exotic 'hoofstock'; others wander the labyrinth of cages at ground level, where they can look a skittish baby camel in the eye. Some of the biggest attractors are the various trailers or side rooms that act as 'warm rooms' for hundreds of caged birds, snakes, reptiles, and tortoises. Baboons, monkeys, and predators, like servals (mid-sized African wild cats) and bears, are also big crowd-pleasers if they are permitted for sale.

The auction space is designed to control these animals: pens, cages, and the auction ring itself are constructed to display and subordinate the animal. Containment and controlled mobility are both integral modes of controlling the animal. In other words, animals are contained in pens, cages, chutes, and trucks at auction in order to keep them captive. But their efficient and controlled mobility is necessary to keep the flow of capital moving through the auction yard. Thus, each piece of fencing is often hinged and moveable, ready to convert to a pen (for containment) or a chute (for efficient movement) (Gillespie 2014). Further, the display of the animal is a crucial part of the commodification process in the front and back of the auction. Before the auction begins, buyers wander through the rear holding pens observing the animals on display, making evaluations and decisions about which animals they will bid on. The space and design of the auction facilitates display and visibility of animals from all angles: for example, the thin, horizontal bar fences permit people to peer inside pens, catwalks are elevated above the holding pens, and stadium-style bleachers give audience members an unobstructed view of the ring from every seat in the auction house.

Subordination is necessary for the animal to circulate as capital. The management and control of animal bodies can be seen in the aggregate – moving animal herds in and out of the auction yard with efficiency – and in the individual animals, like the spider monkey and the cow with barcode #743, whose bodies are subjected to intimate forms of violence, control, and subordination at the auction yard. The auction is also a node in a commodity circuit that more broadly engenders and depends on control at the level of animals' bodily space. In the dairy industry, breeding practices characterized by sexualized violence and a kind of gendered commodification of the animal, driven by efforts to maximize profits and efficiency, have defined the body space of the cow through genetic selection for milk quality, quantity and flavor, body size, reproductive potential, and temperament (Gillespie 2013). In the exotic pet trade, animals' wild lives, in which they move through space freely and engage in complex social practices with their kin, are fundamentally transformed the moment they are captured and transported worldwide to be a caged companion. Often, these animals experience varying degrees of bodily modification, such as having teeth or claws removed, wings clipped, and tracking tags inserted under the skin. Cows used for dairy and exotic pets are thus both subject to profound degrees of spatial control, not only over their movement through confinement but also over the space of their own bodies.

Entangled material and discursive performances at the auction are also central to the animals' commodification. The auctioneer's incessant calls and the interruptions in this hum when the auctioneer speaks to the audience or comments on the animal on display are integral to the discursive construction of the animal-as-commodity. This banter is light and jovial, littered with jokes and laughter that work to normalize the subordinate position of the animal. The lightheartedness surrounding animals' commodification and the violence to which they are subjected highlight the paradox of their valuation: their lives are disposable even as they are central to the circulation of capital. Farmed animal auctions are mundane and animals sell with little fanfare, although some auctions have live bands playing music and most have restaurants where buyers and spectators enjoy a meal of animal-derived foods before and after the sale. At the exotic pet auction, the auction itself is a raucous spectacle. Ring men jokingly attempt to mount bison and make a pretense of using chairs to spar with horned animals; small children ride zebras and camels; and occasionally women enter the ring to hold animals like baby lemurs, kinkajous, and bearded dragons in their cupped hands, lifting them high in the air as they walk around the ring, occasionally bringing the animals down to their faces for a kiss or snuggle.

In different ways, the performance and display at the farmed animal auction and the exotic pet auction are integral to the logic of human dominance exhibited in these spaces. Farmed animal auctions erase the violence of commodification through the everydayness of the auction yard, the jovial nature of farming communities coming together in this place, and the routine efficiency with which they sell the animals. Exotic pet auctions, by contrast, obscure these violent interspecies relations through the spectacular dimensions of the performance, the exotic nature of the animals and their fetishization, and the construction of the auction as a special event (these auctions are held only a few times a year). Both types of sales reveal the auction as a space involved in the reproduction of a capitalist system that renders animals sellable, buyable, breedable, displayable, and, eventually, killable.

The broader political economic conditions under which the auction is made possible are defined by a global system of capitalist exchange that renders 'nature' commodifiable. Animals, as part of this nature, are categorized as resources to be used in the accumulation of capital for human producers and consumers. Auctions are places that reveal the socially constructed and variable nature of economic value, which is determined at auctions by the conditions of the commodity, the conversations between buyers and sellers, and the context in which they are sold (Smith 1989). The collective means by which value is assigned at auctions – performed in front of an audience of tens to thousands – means that they often act as a legitimating force for commodities whose value or status is politically, economically, or ethically uncertain (Smith 1989). For this reason, observing the auction event is to observe the maintenance or remaking of particular commodity forms (Gillespie 2014; Collard forthcoming). In this process of capitalist exchange, the animal is subordinated anew. As the spider monkey enters the auction ring, and is observed, bid on, and sold, her history as a proven breeder and the fact that "her lady parts still work" promise that she will be re-entered into the commodity circuit, involved in the future reproduction of dominant/subordinate human–animal relations.

The ethical and political dimensions of the commodification process can be obscured in the auction space by the perceived 'perfect market' nature of auctions, their operation as either a mundane practice or a spectacle, and the normalization of animal use through various activities and performances at the auction. But auctions are, as various scholars have argued, actually deeply sociopolitical spaces (Smith 1989; Geismar 2001; Garcia-Parpet 2007). This is occasionally evident during concerted political efforts at auctions, such as circulating petitions ensuring people's 'right' to own exotic animals. Even more fundamentally, auctions are – like other economic practices and commodities themselves – formed in and through political, ethical, and social processes. Ethical and political dimensions are particularly important when the commodity is a living being who suffers continually in service to capital accumulation. These animals are "lively commodities," whose lively nature ensures their ability to suffer and simultaneously determines their commodifiability (Collard and Dempsey 2013). In the case of the spider monkey, her value as a lively commodity was tied to her reproductive potential, in spite of the fact that she had missing fingers and a missing eye. Although most exotic pets' lively value is tied to their encounterability more than their reproductive capacity (Collard 2014), the spider monkey's liveliness enabled her to produce new lively (and encounterable) commodities through the reproductive process. The cow with barcode #743 had commodity value when she arrived at auction to be sold for meat, but her value quickly declined when she collapsed in the pen and could not rise, as 'downed' animals are not fit for human consumption. Her body would be rendered into useable byproducts in order to eke out the last bit of capital from her life and death.

As animals continue to be commodified through myriad processes and places under capitalism, the auction is one place where uneven power relations between humans and other animals are readily displayed. The banal violence at the auction is a site through which to understand the invisibilized violence present in humans' everyday use of animals. Auctions are places of capitalist exchange, they are places of encounter between humans and other animals, and they are places where the animals themselves are both displayed and subordinated. Given these multiple dynamics and auctions' inherently spatial nature, it is surprising that geographers have paid so little attention to this space. Economic geographers studying the auction might focus on a spatial political economy of the auction, for example, perhaps noting how humans and animals are embedded in this system of capitalist exchange. Animals' commodity status in this figuring might be noted, but likely not challenged, deferring instead to an analysis that takes animals' commodity status as a given, as has been the trend in economic geography and other sub-disciplines of geography. By contrast, a critical animal geography approach interrogates and challenges the dominant social orders that maintain human–animal hierarchies and perpetuate conditions of animal use. This approach necessarily politicizes entanglements between humans and animals and thinks with ethical and political nuance about the ways animals are subjects of violence and appropriation that often go unquestioned and unchallenged. Critical animal geographies are also, importantly, attentive to the *spaces and places* animals inhabit (Philo and Wilbert 2000). These spaces and places are central to the subordination of animals, as we have seen at the auction: the pens, chutes, cages, and transport trucks are all features of spatial control over animal bodies.

Critical animal geographies are also concerned with the effects for animals of the auction. Behind the hum of the auctioneer's call are the voices of the animals moving through the auction yard. Their bellows, shrieks, bleats, grunts, and squawks echo through the space, a testament of their efforts to communicate the impacts of their commodification. A mother cow and her calf separated in the auction yard, one sold for dairy production and the other sold for veal, call to each other from across the pens. The cage of an African grey parrot is cloaked in a towel, a handwritten sign instructing the towel be left in place; later when it is removed in the auction ring, the parrot hurls his body ferociously around the cage, crashing into its walls until the towel is replaced. A steer escapes from the pens and is chased down and shot, a bull is shocked repeatedly with an electric prod, and a calf is smacked in the face with a paddle.

Critical animal geographies draw attention to the lived experience of the animal, as Philo and Wilbert (2000) early on urged animal geographers to do. Such geographic inquiry aims to understand the embodied experience of domestication, captivity, and the bodily appropriation that is central to the production of animals as food, companions, information, clothing, and so on. Critical animal geographies register and foreground the suffering of the cow with barcode #743 as she struggles to rise from the auction floor and the one-eyed spider monkey curling her partially amputated fingers through the bars of her cage. These realities are not peripheral observations in a geography of the auction yard; rather, they are *central* to an ethically and politically engaged analysis of the fraught entanglements between human and animal lives.

This book aims to bring such a sensitivity and ethico-political charge to animal geography, a sub-discipline of human geography that has, in recent years, experienced a burst of energy and interest. Among the work that has been generated are several important studies that provide signposts and direction for critical inquiry in the sub-discipline. The following section charts this work and elaborates how *Critical Animal Geographies* aims to both build from this foundation and offer new lines of inquiry for the burgeoning field of animal geographies.

Animal geographies: key guides in the 'third wave'

This book is firmly situated within what Julie Urbanik (2012) characterizes as animal geographies' third wave. A key difference between this third wave and two previous twentieth-century swells of geographical interest in animals – the first zoogeographical and the second largely a subset of cultural ecology – concerns how animal geographers imagine the human–animal divide: a mode of ordering both matter and meaning. In this order, humans are not only separate from all other biological life, they are also understood to be superior and transcendent. The human–animal divide can be thought of as what Val Plumwood (1993) calls a dualism, a system in which difference is abstracted into two opposing categories – 'human' and 'animal,' in this case (or 'male' and 'female,' 'culture' and 'nature,' and so on). 'Animal' refers to all beings who (or, according to the language used, 'that') are not human. A caged spider monkey being auctioned off in Missouri, a dying salmon swimming upriver to spawn, a white hen confined with 50,000 others in a confinement shed, Derrida's cat, a beagle in a medical research lab, the matriarch elephant of a herd, a turkey vulture scavenging a decaying rabbit – all are classified 'animal,' a category that masks infinite difference (Derrida 2008). Similarly, the category 'human' conceals the operation of multiple differences, for example along class, race or gender lines.

A dualism does not only abstract difference, it also hierarchizes it. In the human–animal divide, the human is understood as a species of animal that has mastered both its own and external animal nature, and is thought to alone possess capacities such as emotion, reason, tool-use, mourning, lying, lying about lying, subjectivity, a sense of self, and the ability to recognize the self in the mirror, just as a few examples. While humans are technically animals, in the divide, humans are exceptional animals, and so are rarely recognized as animals at all. Thus, another way of thinking about the human–animal divide is as what Donna Haraway (2008) calls "human exceptionalism": the belief in a distinct and superior human figure. As scholars from multiple disciplines have recently argued, the human–animal divide, or human exceptionalism, is not an inherent or pre-given ecological order (e.g., Wolfe 2003; Anderson 2007). It is not ahistorical, apolitical, 'natural' or inevitable. Neither is it universal (Sundberg 2013), although it is dominant. This order is, instead, the product of intersecting social, political, economic, scientific, philosophical, and religious histories and processes – what Giorgio Agamben (2004) calls the "anthropological machine." This machine has all sorts of profound material-discursive effects. It underpins a prejudice we can think of as speciesism, or a disregard for animals on the basis of their species (namely, not being human) and a simultaneous belief in the superiority and mastery of the figure of the human. This perceived absolute difference and hierarchy between humans and animals, enshrined in law, is partly what enables – legally, politically, ethically, socially – animals to be owned as property and exchanged at auctions in ways that we have described. Similarly, this divide has been applied to colonial, racist, and patriarchal projects, licensing the domination of those humans deemed by those in power to be not quite human enough (Plumwood 1993; Wolfe 2003; Haraway 2008).

For early animal geographers, though, the human–animal divide was an untroubled starting place, a foundation, for research. This speciesist divide remained intact and was even further naturalized in bio-geographical studies of wild animal distributions that treated animals as 'natural objects' separate from humans (e.g., Newbigin 1913; see Urbanik 2012 for a summary of first-wave animal geography). Subsequent cultural ecology studies, which retained an interest in animals' spatial distribution, turned their focus to domesticated animals like cows. This work went to greater lengths to understand the cultural significance of animals for human life, and the effects of domestication for landscapes and human–environment relationships. Nevertheless, such research did not extend cultural capacities to animals, and did not consider the divide between the categories 'human' and 'animal' to be itself highly cultural (see Baldwin 1987 and Urbanik 2012 for summaries of animal geographies' second wave). As Jennifer Wolch and Jody Emel (1998) observe, animals did not figure as subjects in these early geographic forays into animal distributions and human–animal relations. Animals were part of an undifferentiated, non-sentient, capital-N Nature: a mass of nonhuman biological life that existed in dualistic relation with human life.

In stark contrast, the past two decades of work in animal geographies have tackled the human–animal divide itself, examining its histories, its social construction, and its effects for human and animal life. As a corollary, third-wave animal geographers recognize animal subjectivity (Emel *et al.* 2002); they recognize that there is no clear-cut dividing line between reasoning, emotional, agential, and self-aware humans (subjects) on one side and passive, mechanistic, 'dumb' animals (objects) on the other. Animals are understood to be not only vastly different across species, but also within species, as individuals with their own social networks and histories. Animals are also understood to be actors – or beings that have the capacity to direct change – shaping politics, culture, social life, economics, and the construction of space and knowledge (e.g., Hobson 2007; Lorimer 2007; Johnston 2008; Sundberg 2011; Dempsey 2010; Collard 2012; Barua 2013; see Buller 2013 for an overview of recent work). Influenced by work in posthumanism and feminist science studies (e.g., Barad 2007) and actor-network theory (e.g., Latour 2005), animal geographers often conceive of this agential capacity relationally. That is, animals are understood to act in concert with other beings and things, including humans.

One of the major undertakings of third-wave animal geography has therefore been to point to and cultivate shared space and networks. "Our political task," write early third-wave animal geographers Wolch and Emel (1998: xii), "is the creation of many forms of shared space." Subsequent animal geographers have explored just that, pointing to human–animal coexistence in urban space, for example (Philo 1998; Wolch 2002; Hovorka 2008), or the space of the home and garden (Power 2009; Ginn 2013), and even our own bodies (Hird 2010). The objective of this work has been to identify the minglings of human–animal life, even in spaces previously thought to be the domain of humans alone, and to emphasize the ethical possibilities found in paying attention to animal others within these shared spaces. Much of the latter in particular has built from Haraway's (2003, 2008: 15) work, which suggests that it is in "face-to-face meetings" with multispecies others that we might learn how to get "on together with some grace."

Now, having acknowledged the multitude of shared spaces – indeed the impossibility of anything but – we are confronted with another, perhaps more urgent, question. How can we more *justly* share space? Identifying relations has been a critical political task, forcing us to confront the infinite relations that constitute us and to which we are therefore bound. But as Judith Butler (2004: 24) writes, "we are not only constituted by our relations but also dispossessed by them." Critical scholars must be attuned to these acts of dispossession, to winners and losers, to the unevenness of power, wealth, and resources within relational networks. This means paying particular attention to capitalism and colonialism as dominant political-economic modes. As geographers such as Michael Watts (2000), Heidi Nast (2006a, 2006b) and Laura Hudson (2011) have shown, animal and human dispossession, oppression, and exploitation are entwined in historical and contemporary capitalism, and occur through enclosure and commodification that can proceed through forging new intimacies as much as severing ties. Understanding these processes, and what they mean for how we work towards justice across species, is a key political project of this edited volume. We argue that in some cases, it may be that multispecies justice is found in spaces where distance is sought after more than proximity, difference more than similarity. This may be a world in which the project is not to enlarge the circle of beings that matter, to bring animals "into the circle of morality and subjectivity" (Wolch and Emel 1998: xii), but rather to re-examine the practice of circling itself. For what is at the heart of the circle? What or who does the circle defend? If it is a figure of the human that is exceptional and entitled, then perhaps this needs to be rethought.

This refiguring of human–animal relations is part of radically changing our material encounters with other animals and our conceptualization of certain species. It may mean the cessation of intervening in animal lives in the case of breeding, culling, and exercising other forms of intimate control over animal bodies. This new order might mean letting go of our attachments to preserving certain species, like domesticated turkeys, pigs, chickens, and cows whose existence is perpetuated by our continued exploitation of them and whose history of human selection for certain traits in breeding has often been the cause of their suffering. It may mean radically revising (and likely doing away with) our categorizations of animal species – as food, pets, research subjects, pests – that also work to subordinate certain species and reify others. It certainly means destabilizing speciesism and redefining how we think about intelligence, cognition, emotional depth, and other physical, emotional, and psychological capacities. In other words, 'human' would no longer be the category against which all other species are measured and come up short.

The project and outline of the book

Our aim with this text, then, is to define and explore distinctly *critical* animal geographies. We draw on the tradition of critical geography to inform our approach. Critical geography questions taken-for-granted spatial orders, or what geographers call spatial imaginaries. Critical geography is especially attentive to the power relations that particular spatial orientations and imaginaries reveal (e.g., the space of the cage – see Karen Morin's chapter in this volume). Critical geography also explores how violence depends in part on specific spatial arrangements: proximities and distances, spatial demarcations such as borders or pens, and controlled movements through space. Critical animal geographies address these questions in the context of human–animal relations. And as critical animal geographers we are not only interested in the animal condition, but also in thinking about how 'the human' and 'the animal' are themselves social and spatial orders.

This chapter's opening analysis of the auction yard illustrates the kinds of questions we see as central to critical animal geographies. What are the embodied effects for animals of humans' use? How are the places and spaces of animal commodification and appropriation, including the space of the body, involved in their continued subjugation (e.g., Boyd 2001; Watts 2004; Gillespie 2013)? How does an attention to forms of animal oppression help us to understand the material effects of exploitative institutions, like capitalism, more broadly? What does a more compassionate and ethically and politically attuned way of being in the world with other animals look like? And how can we get there? These kinds of questions define the larger project we aim to lay out for critical animal geographers going forward.

As a contribution to the critical animal geographies project, this book is organized in three thematic sections: politics, intersections, and hierarchies. These themes, which are elaborated below, emerged organically out of the contributions offered by the authors in this book. We consider them to be key lenses for the kinds of interventions critical animal geographers are making at the turn of the twenty-first century. In this current moment, we take as our starting point that animals' lives and deaths are deeply political, that they cannot be separated from intersecting forms of justice (social, political, environmental), and that hierarchical orders are a primary source of animals' subordination. While we have organized the book into these three distinct sections, the authors' analyses address these themes simultaneously and, as such, considerations of politics, intersections, and hierarchies thread through the entirety of the text.

Politics

As Kersty Hobson (2007: 251) writes, "animals are already subjects of, and subject to, political practices." In this vein, each of the chapters in this first section works toward and wrestles with the contours of a more *radical politics* that might accompany and reflect the intellectual and ethical recognition of how animals matter and how they are affected collectively and as individuals. In doing so, this section evaluates existing approaches to critical animal politics and advances alternative political formations and strategies for critical animal geographers. Each of the three chapters in this section brings together different strands of critical thought – anarchist, posthuman, feminist – with analyses of how animal politics are enacted. Collectively the chapters enliven several key debates within animal politics, many of which are also taken up in other chapters in the book. At what scale should politics be enacted: the individual animal or the collective? At what affective register should political strategy occur: through a politics of sight, or through embodied or gustatory practice? The limits and possibilities of these various approaches are the focus of this section's chapters.

The politics section opens with Richard White's challenge to animal studies scholars, especially animal geographers, to more explicitly address and seek to transform animals' widespread exploitation, oppression and domination. To guide such a critical orientation, White suggests animal geographers look to both critical animal studies (CAS) and anarchist praxis, in particular their emphasis on the 'animal condition,' or the situation for actual living and dying animals. In return, critical animal geographers can offer to CAS a spatial orientation, a commitment to illuminating the invisibility and logic of particular animal spaces. To ground this work, though, White argues that critical animal geographers must find a means of engaging with activist struggles. Eva Giraud's chapter takes up precisely this task. She examines existing political movements that put White's recommended approach – anarchism and critical animal studies – into action, focused on vegan praxis in food movements. Giraud counters some of the critiques that have been made of CAS with an engagement with posthumanist theory, and at the same time advances a picture of what posthuman theoretical approaches might look like in radical and political practice. Grappling with how precisely to articulate posthumanist theory with activist praxis, a coupling she sees as central to critical animal geographies, Giraud's chapter points to several tactics activists can undertake to remain in step with posthumanist ideas. She also finds that activist praxis can and does have a strong geographical orientation, shedding light on invisible spaces and practices, such as grocery store waste, in the manner White advocates.

This objective of making visible is part of a broader politics of sight that is gaining popularity in contemporary animal politics. As the now common adage that opens Claire Rasmussen's chapter goes, "if slaughterhouses had glass walls, we'd all be vegetarians." Rasmussen's chapter troubles this assumption, though, evaluating and complicating the strategy of making visible and the assumption that seeing can change people's ethical sensibilities. To construct her analysis, Rasmussen places the legal debates around U.S. 'ag-gag' laws (state bans on filming and photography in animal production facilities) alongside another legal struggle occurring in the U.S. over the ability to see animal violence, in this case, to see 'animal crush' videos, or films in which animals are intentionally harmed, usually for sexual gratification. In doing so, Rasmussen points to differentiated affective responses to seeing animal violence. She ultimately argues that an ethical equation that rests on seeing-equals-knowing-equals-recognition remains caught within a liberal regime of rights that is not capable of challenging animal violence because as a system it relies on that violence.

Intersections

Intersectional analyses are also a key feature of critical animal geographies. Questions of justice for animals cannot be divorced from other forms of justice (for humans, for the environment) in geographical analyses. Hierarchies of power operate across human and animal social groups to subordinate certain lives and bodies based on overlapping axes of exploitation including race, gender, sexuality, class, ability, and species. And while the specific impacts and enactments of these power hierarchies unfold in particular ways for different bodies, the structures that maintain these modes of subordination function similarly. Therefore, while we recognize the excellent critiques of intersectionality made in recent years (e.g., Puar 2007), we still find, as do many ecofeminists and critical animal studies scholars, that it is an indispensable tool for thinking about human–animal relations. While not all of the chapters in the 'intersections' section take up an explicitly intersectional approach, all feature research purposively located where the exertion of violence and power against humans meets that exerted against animals. These chapters all consider in one way or another connections between animal and human oppression, with attention to race, gender, class, species, and environment. Building from a well-established body of work in animal geography that explores how animals and speciesism are enrolled in the formation of racialized and gendered bodies and identities, these chapters highlight techniques of power and management regimes that cross human and animal lines. All of these chapters track, in one way or another, the *production* of racialized and animalized subjects – the caged animal and the caged prisoner, the colonized, the enslaved, the worker, and so on. In doing so, the authors actively contest the idea that these subject positions are innate, pre-given, or essential.

Karen Morin's opening chapter, for example, highlights the production of two different caged bodies and spaces: animals on display at the zoo and humans incarcerated in the prison. Examining disciplinary regimes and mandates at work in the securing of both caged bodies and spaces, she shows how both prisoner and caged animal are produced as animalized subjects – dangerous, wild, savage – arguing that spatial tactics of controlled (in)visibility are an especially integral part of caged bodies' animalization and objectification. Tracing the evolution of both zoo and prison cages in recent history, Morin draws out parallels and divergences within these histories and future trajectories, in which zoos appear poised to be a more likely candidate for extinction than prisons. She highlights what can be gained by thinking across these experiences, especially for those who seek to contest caging in either circumstance. John Joyce, Joe Nevins, and Jill Schneiderman also focus their analysis on a form of enclosure: the Hudson Valley Foie Gras (HVFG) facility in New York State. Engaging a political economic and ecofeminist theoretical approach, their objective is to understand how the abuse and commodification of human and nonhuman beings at HVFG might be "dynamically interrelated," something the authors maintain activists/critics have failed to do. To do so, Joyce, Nevins, and Schneiderman interpret HVFG as a dense nexus within international material flows of ducks and laborers and circulating ideologies of speciesism, racism, nation, gender, and citizenship. They find that the foie gras industry materially and symbolically "deadens" ducks and human laborers, relying on "intersecting trajectories" of mobility and immobility for humans and ducks.

Turning attention away from domesticated and caged animals, Anastasia Yarbrough provides a detailed critique of historical and contemporary wildlife management in the U.S., showing how racism and colonialism play an enduring and potent role. Wildlife management has, from its inception, been about protecting some people's – namely wealthy, white male elites' – rights to 'harvest' wild animal resources; it was never about serving the interests of wild animals. Problems with management regimes today are in step with this history: from suspect hiring practices to selective recognition of stakeholders (largely defined as those who 'use' animals – i.e., hunt them), U.S. wildlife management fails, Yarbrough argues, to recognize people of color, indigenous people, and wild animals as "legitimate subjects in a community of biocultural diversity." Interspecies collaboration and environmental justice approaches offer a more promising alternative, Yarbrough suggests in conclusion. Taking a closer look at some of the same dynamics of U.S. colonialism and racism that Yarborough's chapter illuminates, Heidi Nast uses primary sources to follow early pit bull fighters, predominantly white working-class coal miners and manufacturers in the UK, who introduced the dogs into the U.S. in the 1800s. Once in the U.S., pit bulls were put to various uses within colonial expansion, including tracking runaway slaves. White settlers also fought their dogs, only joined in the fighting pit by black men after civil war emancipation. In ensuing decades, black men were depicted as inept at dog fighting, an activity associated with whiteness. White supremacy was thus maintained through a disciplining of what was the 'right' kind of violence humans could commit against animals and against each other, a mode of racialization that continues today, as Nast shows through her concluding reflection on contemporary portrayals of dog fighting that frame it as largely the domain of men of color.

Hierarchies

Critical Animal Geographies takes as a central starting point that entrenched hierarchies between humans and other animals are a defining feature of our relationships. Although the human exceptionalism of Western humanism is by no means universal (see Sundberg 2013), an anthropocentric species hierarchy, with humans positioned at the top, underpins the world's dominant political economies and ecologies, and justifies acts of domination and appropriation of other species. These power-laden categorizations materialize on the ground in places like the auction yard, where the design, maintenance, and general functioning of the auction rely on the subordinate animal. Our attention to hierarchy threads through this text, culminating in a final section of the book in which chapters take on the complicated and ambivalent task of how to think and act responsibly and ethically in a world of species hierarchies that are constantly being remade and reconfigured. Negotiating species difference and differential power is ever-present and ongoing. Proceeding from this point, each chapter in this section considers encounters or institutions in which species hierarchies are actively negotiated, challenged, transgressed or restructured. They point on the one hand to the dynamism of species hierarchies, and on the other to enduring orders. As such, each chapter is concerned with ethics, with how we respond to animals and how we might respond *better*, or more gently. All of the chapters ask this question in high-stakes situations where beings are dying, killing, and being killed, and where the terrain of killability is dynamic and contested.

Gwendolyn Blue and Shelley Alexander's study of urban coyotes carefully tracks such a zone of contested killability. Coyotes, like other large predators, have the potential to upset imagined species hierarchies, inverting human dominance. So how do we respond to an animal who may regard us as food? Blue and Alexander offer a "gastro-ethical" approach to eating well, following Derrida and Haraway. Disrupting the human exceptionalist notion that humans are atop, but not part of, the food chain, this approach centrally acknowledges that we all eat and that killing and eating are entangled. This approach does not aim to replace tricky ethical decisions with a universal ethical calculus, but rather provides the ground for working out ethical dilemmas that are always ongoing and embodied. Like Blue and Alexander, Jody Emel, Connie Johnston, and Elisabeth Stoddard explore what a flatter species hierarchy might look like with respect to eating. If Blue and Alexander are largely focused on the political and ethical struggles that accompany decisions to kill wild animals (or make them killable), Emel, Johnston, and Stoddard primarily direct their attention to the conditions of life for domesticated animals, in their case pigs who will eventually be killed to become meat. They centrally grapple with whether "livelier livelihoods" might be facilitated for these animals, and if so, how. Guided by a scientific approach to animal welfare, the authors' own assessment of two modes of pig farming, CAFO and permaculture, leads them to conclude that generally the former tends to foreclose animal agency while the latter enables it.

Finally, Irus Braverman examines an institution and practice within which species hierarchies are refashioned through the formulation and constant revision of endangered species lists. The conservationist and scientific authors of these lists are engaged in setting the parameters (whose success is not guaranteed) used to decide which animals are more or less at risk of extinction. Through interviews and archival research, Braverman documents the details of this decision-making and its consequences. Arguing that threatened species lists are affirmative biopolitical technologies, she shows how the list becomes a way not only to affirm and justify those animal lives most important to save, but also, on the flip side of the coin, to manage death at the level of the population, establishing those lives which remain killable. Through her attention to this aggregate-scale deployment of power-over-life, Braverman draws attention to the importance of interrogating and challenging the category of 'species' as a unit of conservation governance.

As a whole, *Critical Animal Geographies* intervenes into the exciting explosion of interest in animals within geography and beyond with the aim of orienting future research. This hopeful future is traced in the conclusion to the volume, which evaluates existing gaps and needed lines of inquiry. All of the chapters in the book are united in their challenge to dominant modes of human–animal relations, and in envisioning more ethical and compassionate ways of being with, and letting be, multispecies others. We hope that the contributions herein will help to chart a course toward more critical, radical, and provocative animal geographies.

References

Agamben, G 2004, *The open: man and animal*, Stanford University Press, Stanford.
Anderson, K 2007, *Race and the crisis of humanism*, Routledge, New York.
Baldwin, J 1987, 'Research themes in the cultural geography of domesticated animals, 1974–1987,' *Journal of Cultural Geography*, vol. 7, pp. 3–18.
Barad, K 2007, *Meeting the universe halfway: Quantum physics and the entanglement of matter and meaning*, Duke University Press, Durham, NC.
Barua, M 2013, 'Volatile ecologies: towards a material politics of human–animal relations,' *Environment and Planning A*, vol. 46, pp. 1462–1478.
Boyd, W 2001, 'Making meat: science, technology, and American poultry production,' *Technology and Culture*, vol. 42, pp. 631–664.
Buller, H 2013, 'Animal geographies I,' *Progress in Human Geography*, vol. 38 (2), pp. 308–318.
Butler, J 2004, *Precarious life: the powers of mourning and violence*, Verso, Brooklyn.
Collard, R-C 2012, 'Cougar-human entanglements and the biopolitical un/making of safe space,' *Environment and Planning D: Society and Space*, vol. 30, pp. 23–42.
Collard, R-C 2014, 'Putting animals back together, taking commodities apart,' *Annals of the Association of American Geographers*, vol. 104, pp. 151–165.
Collard, R-C forthcoming, *Animal traffic*, Duke University Press, Durham, NC.
Collard, R-C and Dempsey, J 2013, 'Life for sale? The politics of lively commodities,' *Environment and Planning A*, vol. 45, pp. 2682–2699.
Dempsey, J 2010, 'Tracking grizzly bears in British Columbia's environmental politics,' *Environment and Planning A*, vol. 42, pp. 1138–1156.
Derrida, J 2008, *The animal that therefore I am*, Fordham University Press, New York.
Emel, J, Wilbert, C, and Wolch, J 2002, 'Animal geographies,' *Society and Animals*, vol. 10, pp. 407–412.
Garcia-Parpet, M 2007, 'The social construction of a perfect market: the strawberry auction at Fontaines-en-Sologne,' in C MacKenzie, F Muiesa, and L Siu (eds), *Do economists make markets? On the performativity of markets*, pp. 20–53, Princeton University Press.
Geismar, H 2001, 'What's in a Price? An Ethnography of Tribal Art at Auction,' *Journal of Material Culture*, vol. 6, pp. 25–47.
Gillespie, K 2013, 'Sexualized violence and the gendered commodification of the animal body in Pacific Northwest US dairy production,' *Gender, Place and Culture*, early view.
Gillespie, K 2014, 'Reproducing dairy: Embodied animals and the institution of animal agriculture,' Doctoral Dissertation, University of Washington.
Ginn, F 2013, 'Sticky lives: slugs, detachment and more-than-human ethics in the garden,' *Transactions of the Institute of British Geographers*, early view.
Haraway, D 2003, *The Companion Species Manifesto*, Prickly Paradigm Press, Chicago.
Haraway, D 2008, *When species meet*, University of Minnesota Press, Minneapolis.
Hird, M 2010, 'Meeting with the microcosmos,' *Environment and Planning D: Society and Space*, vol. 28, pp. 26–29.

Hobson, K 2007, 'Political animals? On animals as subjects in an enlarged political geography,' *Political Geography*, vol. 26, pp. 250–267.

Hovorka, A 2008, 'Transspecies urban theory: chickens in an African city,' *Cultural Geographies*, vol. 15, pp. 95–117.

Hudson, L 2011, 'A species of thought: bare life and animal being,' *Antipode*, vol. 43, pp. 1659–1678.

Johnston, C 2008, 'Beyond the clearing: towards a dwelt animal geography,' *Progress in Human Geography*, vol. 32, pp. 633–649.

Latour, B 2005, *Reassembling the social: An introduction to actor-network-theory*, Oxford University Press, Oxford and New York.

Lorimer, J 2007, 'Nonhuman charisma,' *Environment and Planning D: Society and Space*, vol. 25, pp. 911–932.

Nast, H 2006a, 'Critical pet studies?,' *Antipode*, vol. 38, pp. 894–906.

Nast, H 2006b, 'Loving… whatever: alienation, neoliberalism and pet-love in the twenty-first century,' *ACME: An International E-Journal for Critical Geographies*, vol. 5, pp. 300–327.

Newbigin, M 1913, *Animal geography: the faunas of the natural regions of the globe*, Clarendon, Oxford.

Philo, C 1998, 'Animals, geography, and the city: notes on inclusions and exclusions,' in J Wolch and J Emel (eds), *Animal geographies: Place, politics and identity in the nature-culture borderlands*, pp. 51–70, Verso, London.

Philo, C and Wilbert, C 2000, 'Introduction,' in C Philo and C Wilbert (eds), *Animal spaces, beastly places*, pp. 1–34, Routledge, London.

Plumwood, V 1993, *Feminism and the mastery of nature*, Routledge, New York.

Power, E 2009, 'Border-processes and homemaking: encounters with possums in suburban Australian homes,' *Cultural Geographies*, vol. 16, pp. 29–54.

Puar, J 2007, *Terrorist assemblages: Homonationalism in queer times*, Duke University Press, London.

Smith, C 1989, *Auctions: The social construction of value*, University of California Press, Berkeley.

Sundberg, J 2011, 'Diabolic Caminos in the desert and cat fights on the Río: a posthumanist political ecology of boundary enforcement in the United States–Mexico borderlands,' *Annals of the Association of American Geographers*, vol. 101, pp. 318–336.

Sundberg, J 2013, 'Decolonizing posthumanist geographies,' *Cultural Geographies*, vol. 21, pp. 33–47.

Urbanik, J 2012, *Placing animals: An introduction to the geography of human-animal relations*, Rowman & Littlefield, Lanham, MD.

Watts, M 2000, 'Afterword: enclosure,' in C Philo and C Wilbert (eds), *Animal spaces, beastly places: New geographies of human-animal relations*, pp. 291–301, Routledge, London.

Watts, M 2004, 'Are hogs like chickens? Enclosure and mechanization in two "white meat" filières,' in A Hughes and S Reimer (eds), *Geographies of commodity chains*, pp. 39–62. Routledge, London.

Wolch, J 2002, 'Anima urbis,' *Progress in Human Geography*, vol. 26, pp. 721–742.

Wolch, J and Emel, J 1998, 'Preface,' in J Wolch and J Emel (eds), *Animal geographies: Place, politics and identity in the nature–culture borderlands*, pp. x–xxii. Verso, New York.

Wolfe, C 2003, *Animal rites: American culture, the discourse of species, and posthumanist theory*, University of Chicago Press.

Part I
Politics

2   Animal geographies, anarchist praxis, and critical animal studies

Richard J. White

Anarchism is the purity of rebellion. A pig who struggles wildly and rends that air with his cries while he is held to be slaughtered, and a baby who kicks and screams when, wanting warmth and his mother's breast, he is made to wait in the cold – these are two samples of natural rebellion. Natural rebellion always inspires either deep sympathy and identification with the rebelling creative, or a stiffening of the heart and an activation of aggressive-defensive mechanisms to silence an accusing truth. The truth is that each living being is an end in itself; that nothing gives a being the right to make another a mere instrument of his purposes. The rebel against authority holds to this truth in everything that concerns him and recognizes no other judge than himself [*sic*].

(Baldelli 1971: 17)

Introduction

In the late 1990s, as a revitalized interest in (nonhuman) animal geographies was gaining new momentum and traction in the discipline, Wolch and Emel (1998: xi) called for geographers to *critically* engage with the brutal and deceptive geographies of violence that frame many ordinary, everyday human–animal relations in society:

The plight of animals worldwide has never been more serious than it is today. Each year, by the billions, animals are killed in factory farms; poisoned by toxic pollutants and waste; driven from their homes by logging, mining, agriculture, and urbanization; dissected, re-engineered, and used as spare body-parts; and kept in captivity and servitude to be discarded as soon as their utility to people has waned. The reality is mostly obscured by the progressive elimination of animals from everyday human experience, and by the creation of a thin veneer of civility surrounding human–animal relations, embodied largely by language tricks, isolation of death camps, and food preparation routines that artfully disguise the true origins of flesh-food. Despite the efforts made to minimize human awareness of animal lives and fates, however, the brutality of human domination over the animal world and the catastrophic consequences of such dominionism are everywhere evident.

The intervening fifteen years, evidenced by the remarkable number of papers, articles, books, conferences, seminars, workshops, modules, courses, and study groups, has seen broader scholarly interest in animal geographies expand considerably. The effect of this on the wider discipline, as Buller (2013a: 308) argues, is akin to:

> A gathering swarm, a swelling herd, a flock or a vast shoal; animal geographies … [has] become an increasingly present, dynamic and potentially innovative subfield of geography (to the point at which some hesitate now to refer to a solely 'human geography').

Similarly, across the broader social sciences, recent decades have served witness to a rapid expansion of interest focused on bringing nonhuman animals out of the shadows of marginality. Diverse approaches are now found in various academic fields such as animal studies (AS), human–animal studies (HAS), and anthrozoology (for a breakdown of these different nomenclatures see Taylor and Twine 2014). Unfortunately a great deal of this research has found comfortable alignment within a largely apolitical, safe, and sanitized discourse, lacking any *critical* commitment to explicitly challenge the exploitation, oppression, and domination that define animal lives (and deaths).

Moreover, strong arguments have been made to suggest that esoteric, conservative, apolitical, amoral, and speciesist approaches have assumed a hegemonic status within mainstream animal studies (see Best 2009). As a reaction to this, as the chapter will argue at greater length, an explicit field dedicated to a Critical Animal Studies (CAS) approach emerged. For Pedersen and Stănescu (2012: ix) a critical approach to animal studies shifts the perspective of enquiry from the animal 'question' to the animal 'condition.' Having much in common with the animal plight that Wolch and Emel (1998) captured, the animal condition is defined as "the actual life situation of most nonhuman animals in human society and culture, as physically and emotionally experienced with its routine repertoire of violence, deprivation, desperation, agony, apathy, suffering, and death" (Pedersen and Stănescu 2012: ix). While acknowledging the important work that has carved out critical spaces within animal geography, the main aim of this chapter is to impress upon the urgency and need for *more* research within the field of CAS to provide more effective tools and strategies of resistance toward the liberation of nonhuman animals, given that "global institutionalized animal use and abuse shows no sign of decline" (ibid.). Within contemporary animal geographies, I want to suggest that many important *liberatory* and critical lines of flight will be created by explicitly engaging anarchist theory and practice. Indeed, at a time when anarchist praxis is blossoming within wider (radical) geography (see Springer *et al.* 2012; Springer 2013) this intervention is particularly timely. Before focusing in more detail on what anarchism and the guiding principles of CAS (themselves influenced by anarchist praxis) have to potentially inform and inspire future directions for animal geographies, it is important to recognize already existing critical dimensions within contemporary animal geographies.

Recognizing the critical in contemporary animal geographies

Emel *et al.* (2002: 407) acknowledge the way in which animal geographies has begun to interrogate human and nonhuman animal relations, and call for such critical engagement to expand further:

> Geography, as a discipline, has provided significant leadership in explicating the history and cultural construction of human and nonhuman animal relations, as well as their gendered and racialized character and their economic embeddedness. This work must continue. There are wide areas of barely touched terrain in comparative cultural analyses, economies of animal bodies, and the geographical history of human–animal relations that need articulation and examination.

Reviewing the animal geographies literature reveals an ongoing critical pulse, where the focus on nonhuman animals is consistent with Johnston's (2008: 633) claim that "the field of animal geographies has arisen in response to our political and ethical responsibilities to the species who share our planet." For example, geographers have sought to disturb and reconfigure dominant human–animal relations through exploring a myriad of ethical dimensions (e.g., Jones 2000; Matless 1994; Roe 2010); engaging with everyday interspecies relationships (e.g., Braun 2005; Tipper 2011; Urbanik and Morgan 2013); rethinking specific relationships with other animals (e.g., Nast 2006); understanding the co-constitutive ways in which nonhuman animals make and shape (urban) places (e.g., Wolch 2002); calling into question representations of 'the wild' and 'wildlife' (e.g., Whatmore and Thorne 1998); examining gender difference in human–animal interactions (e.g., Herzog 2007); and calling for nonhuman animals to be included in an enlarged political geography (e.g., Hobson 2007). More recently, intersectionality and common natures of oppression have begun to be explored. For example, through her work on gendered commodification and the production of dairy and 'meat,' Gillespie (2013: 2) argues that:

> Understanding this commodification is important both for the sake of the individual animals laboring and dying within the industry and for the more extensive project of uncovering the consequences of gendered commodification of all bodies – nonhuman and human – and the violent power structures to which they are subjected.

In many ways, this research has been genuinely liberating and progressive in problematizing and collapsing entrenched binaries concerning human–animal relations, nature-culture, and exposing the anthropocentric, speciesist and humanist roots that have informed representations of 'the animal.'

Another significant contribution by animal geographers can be seen in the desire to reject the undifferentiated and impersonal collective (*species, herds, 'farmed' animals*) and emphasize the individual and unique natures of those sentient beings which are all "subjects of a life" (Regan 2001: 17). This powerful focus on language is a significant step forward to prompt new interspecies connections, by provoking more radical ethical and political questions about our relations with other animals. For example, the (ethical) importance of transgressing the metaphoric collective is exemplified by Jones (2000: 281), who argues that:

> The moving away from 'face-to-face' positioning of non-humans to making them 'faceless' things must contribute to the cruelty many face today.... Any possible switch from relating to non-human others as collectives to relating to them instead as individuals has profound implications for how we live on this planet, and may have a significant narrative for the future.

From a moral perspective, the individualization of animals as opposed to their aggregation also helps undermine the view that sentient pain and suffering can be meaningfully aggregated. In contrast to utilitarian moral theory (e.g., Singer 1995), the argument that pain (or pleasure) cannot be meaningfully quantified and traded between individuals as only individuals experience pain (or pleasure) forms the core principle of Ryder's (2010: 402) moral theory of "Painism." In this approach the focus of (moral) concern is one which should be 'victim centered' and thus never lose sight of the need to concentrate on the "conscious experience of individuals" (Ryder 1998: 269). Importantly, creating new individualized spaces allows new depths, relationships, connections, and understanding to emerge. As Masson (2004: x) argues:

> Animals have a past, a story, a biography. They have histories. Mink and bears, elephants and dolphins, pigs and chickens, cats and dogs: each is a unique somebody, not a disposable something.
> Think of the many implications: animals have mothers and fathers, often siblings, friendships, a childhood, youth, maturity. They go through life cycles much the way humans do ... their lives can go better or worse for them, whether or not anyone else cares about this.

However, *despite* these promising interventions, and critical contributions within animal geography, the fact remains that *more* nonhuman animals are subject to routine repertoires of violence than ever before. The nonhuman animal plight – in terms of the number of nonhuman animals involved – is more extensive, and more serious now than at any other time before. Focusing on the farming of nonhuman animals alone, that practice which "has long been, and continues to be, the most significant social formation of human–animal relations" (Calvo 2008: 32) – the number of nonhuman animals (ab)used is to all intents and purposes unimaginable. Buller (2013b: 157) draws attention to the fact that "the vast majority of the 24 or so billion terrestrial farm animals that are kept and grown for human and other consumption at any one time do so on farms, with an increasing proportion of them on large scale, industrial farm units." Elsewhere, Mitchell (2011: 38) observes that:

Worldwide, approximately 55,000,000,000 land-based nonhumans are killed every year in the farming industry (United Nations Food and Agriculture Organization, 2010). *This is over 150 million individuals each day....* Except for a very tiny minority, all the nonhumans in the industry will meet with a violent death at a relatively young age; all will have been confined during their lives; many will have been mutilated; numerous females will have been repeatedly made pregnant but their young taken away shortly after birth; family structures will have been destroyed.

(Italics added)

Thus, *despite* these critical and potentially transformative spaces that can be found within animal geographies – which contribute significantly to painting a richer, more detailed emotional, psychological, and mental tapestry of nonhuman animals, their capacities for pain and pleasure, love, friendship, companionship – the sorry truth remains that *more* nonhuman animals are subjected to systematic repertoires of violence than at any other time in history. Reflecting on the mass slaughter of farmed animals in response to the non-lethal foot-and-mouth disease in England in 2001, Scully (2002: x) observed that:

in a strange way mankind [*sic*] does seem to be growing more sentimental about animals, and also more ruthless. No age has ever been more solicitous to animals, more curious and caring. Yet no age has ever inflicted upon animals such massive punishments with such complete disregard, as witness scenes to be found on any given day at any modern industrial farm.

Whatever important counter-hegemonic spaces, and strategies of resistance emerging from animal geographies are, they remain insufficient in challenging and transforming these increasingly dominant patterns of violence and dominion. Thus, how can animal geography be pushed forward in new, ever critical and transformative directions to better challenge the oppressive systems of nonhuman animal domination that are becoming ever *more* pervasive and entrenched in society? There is certainly much to be gained by increasing the visibility of those critical pulses evident within animal geography, while also extending these purposefully to inform new tools of resistance that can begin to better confront the (intersectional) systems of dominion that define human and nonhuman animal relations. At the very least if more animal geographers could commit themselves (and their work) through embracing a critical animal praxis that works within a *total liberation* framework, one which challenges *all* forms of domination and exploitation that concern human, nonhuman animals and the Earth, this would be a tremendously important development. As a means of encouraging and stimulating such a future, this chapter highlights how a willingness to engage with anarchist praxis, and Critical Animal Studies (CAS), offers many important ways to think about and, more importantly, actively engage with addressing the animal condition, and transforming this where possible from one of abject misery and suffering, toward peace, pleasure, hope, and fulfilment.

Animal geographies: embracing anarchist lines of flight

Almost forty years have passed since Richard Peet observed that "anarchist theory is a geographical theory" (1975: 43). Yet, only very recently has geography experienced an exciting and long overdue (re)turn to anarchist theory and practice to inspire (human) geographers to challenge and push beyond the existing frontiers of geographical knowledge (Springer *et al.* 2012). What is understood by anarchism in this context? In stark contrast to the deep and introspective theorizing that has accompanied other radical geography approaches, "anarchists have always insisted on the priority of life and action to theory and system" (Weick 1971: 10). Inevitably, such openness has encouraged a rich and diverse range of anarchism to flourish (including collectivist anarchism, anarcho-syndicalism, and more individualist types of anarchism). Highlighting the common bonds that unite all anarchist praxis, Springer (2012: 1606) cites an unequivocal rejection of "all forms of domination, exploitation, and 'archy' (systems of rule)." Indeed, citing *green anarchism* and *anarcha-feminism* as examples, Ward (2004: 3) speaks of anarchists' rejection of all forms of (unjustified) external authority as being attractive to "those who believe that animal liberation is an aspect of human liberation."

Certainly in the history of anarchism, some of the most prominent advocates of, and influences over, anarchist praxis have demonstrated a radical commitment to challenge complex oppressive systems of intersectional domination that limit human and nonhuman integrity and freedom. Emma Goldman (2005: 41–42), for example, when paying tribute to the American anarchist Voltairine de Cleyre, wrote:

> Fortunately, the great of the world cannot be weighed in numbers and scales; their worth lies in the meaning and purpose they give to existence, and Voltairine has undoubtedly enriched life with meaning and given sublime idealism as its purpose. But, as a study of human complexities she offers rich material. The woman who consecrated herself to the service of the submerged, actually experiencing poignant agony at the sight of suffering whether of children or dumb [*sic*] animals (she was obsessed by love for the latter and would give shelter and nourishment to every stray cat and dog, even to the extent of breaking with a friend because she objected to her cats invading every corner of the house).

Elsewhere, standing as an inspirational (though oft overlooked) critical animal geographer is the French anarchist Élisée Reclus (1830–1905). What is particularly important to take from Reclus's work, is (1) the appeal he brings to the abuse of nonhuman animals by drawing on emotional registers (beauty, ugliness), not abstract philosophical rights-based concepts; (2) how he draws on personal experience, and invites others to do the same; and (3) his belief that individuals (and their community) can meaningfully begin to confront and challenge systems of oppression through the decisions they make in the here and now (see White and Cudworth 2014). This approach is captured best in the short pamphlet *On Vegetarianism* (Reclus 2013 [1901]). For example, highlighting incidents – barbarous acts – which cast a gloom over his childhood, Reclus recollects:

> I can still see the sow belonging to some peasants who were amateur butchers, the cruellest kind. One of them bled the animal slowly, so that the blood fell drop by drop, for it is said that to make a really good blood sausage, the victim must suffer a great deal. And indeed, she let out a continuous cry, punctuated with childlike moans and desperate, almost human pleas. It seemed as if one were listening to a child.
>
> (2013: 157)

Reclus then goes on to question the ridicule and disgrace that wider society reserves for those people who bestow affection and love to "an animal who loves us!" (ibid.). Yet these powerful interspecies connections of (mutual) kindness and support are also strongly emphasized by another great anarchist geographer, and contemporary of Reclus, Peter Kropotkin. Epitomized in his work *Mutual Aid* (1998 [1902]) Kropotkin emphasizes that, steeped in the struggle for existence that all life faces, a spirit of reciprocity, mutuality, and sociality is by no means exclusive to the human animal. For Kropotkin (1998: 13):

> Whenever I saw animal life in abundance … I saw Mutual Aid and Mutual Support carried on to an extent which made me suspect in it a feature of the greatest importance for the maintenance of life, the preservation of each species, and its further evolution.

Indeed Kropotkin (1998: 27) also predicted that mutual aid will be found beyond the animal kingdom, noting the cooperation evident amongst invertebrates (termites, ants, bees) and suggesting that "we must be prepared to learn some day … facts of unconscious mutual support, even from the life of micro-organisms."

For all the foundational work and insight that these (classical) anarchists brought to encourage a broader (ethical) sensibility and justice for nonhuman animals, unfortunately, despite the *anarchist* commitment to recognize and confront all sites of domination and exploitation, the inclusion of nonhuman animals in recognizing commonalities of oppression that transgress species membership has been conspicuous by its absence. As Torres (2007: 127) notes, "while social anarchism has been at the forefront of challenging many oppressions, most social anarchists have not been very active – either historically or presently – in challenging the human domination of animals." Indeed several influential anarchists have been hostile to the inclusion of nonhuman animals. Murray Bookchin, for example, argues that only human agency can be regarded as "discursive, meaningful, or moral" (1993: 48). Socha (2012: 15), though, forms an excellent (rhetorical) question, which is the basis on which the rest of the chapter builds: "Proper contemplation of anarchist traditions leads to concern for animals. Can a society whose abiding objective is freedom from violence, hierarchy, and oppression confine, slaughter, dominate, eat and wear other sentient creatures?" No, of course, it cannot. In terms of mapping out potential synergies within already existing critical animal geographies, as well as suggesting further anarchist-inspired lines of flight for animal geographies, I wish to draw attention to the field of Critical Animal Studies (CAS), and in particular its guiding principles, many of which have been inspired by anarchist praxis.

In 2001 Steve Best and Anthony J Nocella II cofounded the Centre on Animal Liberation Affairs (CALA), which was renamed the Institute for Critical Animal Studies (ICAS) in 2006 (see Nocella *et al.* 2014; Taylor and Twine 2014). The term 'Critical Animal Studies' (CAS) emerged following conversations and discussion between a range of animal rights/liberation academics and activists. From its very beginning, CAS was *self-consciously* critical, envisaged as a:

> necessary and vital alternative to the insularity, detachment, hypocrisy, and profound limitations of mainstream animal studies [which] utterly fail to confront [nonhuman animals] ... as sentient beings who live and die in the most sadistic, barbaric, and wretched cages of technohell that humanity has been able to devise, the better to exploit them for all they are worth.
>
> (Best *et al.* 2007: 4)

What should also be recognized are the complex boundaries that overlap between different types of animal studies. In seeking to differentiate itself from other more mainstream forms of animal studies, Critical Animal Studies is not a pure/disconnected/elitist project, nor does it seek to erase – or supersede – other progressive forms of critical research that have taken place, and continue to do so. There are many forms of work – critical theory, ecofeminism, and intersectional theory for example – that have strong antecedents and affinity with CAS, and from which there is much still to be learned and debated. As Twine and Taylor (2014: 4) highlight:

> Both animal and feminist politics are similarly targeted against dispassionate, institutionalised scholarship based on a rationalist, liberal interpretation of (hegemonic) masculinity, and both seek to expose and overthrow the routinised and naturalised forms of practice based on oppression and abuses of power, which flows from this. It is this which makes them both explicitly *critical.*

To give form to the spaces upon which CAS would seek to build, and design a list that would be accessible to both academic and activist communities, Best *et al.* (2007: 4) developed the Ten Guiding Principles of Critical Animal Studies (see Figure 2.1).

There are two further principles that CAS has recently highlighted: "(1) to abolish nonhuman animal oppression, exploitation, and murder on college and university campuses, and (2) to provide space and place for the advocacy of all oppressed groups including nonhuman animals" (Nocella *et al.* 2014: xxviii).

1. Pursues interdisciplinary collaborative writing and research in a rich and comprehensive manner that includes perspectives typically ignored by animal studies, such as political economy.
2. Rejects pseudo-objective academic analysis by explicitly clarifying its normative values and political commitments, such that there are no positivist illusions whatsoever that theory is disinterested or writing and research are nonpolitical. [CAS supports] experiential understanding and subjectivity.
3. Eschews narrow academic viewpoints and the debilitating theory-for-theory's-sake position in order to link theory to practice, analysis to politics, and the academy to the community.
4. Advances a holistic understanding of the commonality of oppressions, such that speciesism, sexism, racism, ablism, statism, classism, militarism, and other hierarchical ideologies and institutions are viewed as parts of a larger, interlocking, global system of domination.
5. Rejects apolitical, conservative, and liberal positions in order to advance an anti-capitalist, and, more generally, a radical anti-hierarchical politics. This orientation seeks to dismantle all structures of exploitation, domination, oppression, torture, killing, and power in favor of decentralizing and democratizing society at all levels and on a global basis.
6. Rejects reformist, single-issue, nation-based, legislative, strictly animal interest politics in favor of alliance politics and solidarity with other struggles against oppression and hierarchy.
7. Champions a politics of total liberation which grasps the need for, and the inseparability of, human, nonhuman animal, and Earth liberation and freedom for all in one comprehensive, though diverse, struggle; to quote Martin Luther King Jr.: *"Injustice anywhere is a threat to justice everywhere."*
8. Deconstructs and reconstructs the socially constructed binary oppositions between human and nonhuman animals, a move basic to mainstream animal studies, but also looks to illuminate related dichotomies between culture and nature, civilization and wilderness and other dominator hierarchies to emphasize the historical limits placed upon humanity, nonhuman animals, cultural/political norms, and the liberation of nature as part of a transformative project that seeks to transcend these limits towards greater freedom, peace, and ecological harmony.
9. Openly supports and examines controversial radical politics and strategies used in all kinds of social justice movements, such as those that involve economic sabotage, from boycotts to direct action, toward the goal of peace.
10. Seeks to create openings for constructive critical dialogue on issues relevant to Critical Animal Studies across a wide range of academic groups; citizens and grassroots activists; the staffs of policy and social service organizations; and people in private, public, and nonprofit sectors. Through – and only through – new paradigms of ecopedagogy, bridge-building with other social movements, and a solidarity-based alliance politics, it is possible to build the new forms of consciousness, knowledge, social institutions that are necessary to dissolve the hierarchical society that has enslaved this planet for the last 10,000 years.

Figure 2.1 The ten guiding principles of critical animal studies (Best *et al.* 2007: 4).

Viewed both individually and collectively many of these principles, with their opposition to all forms of domination and authoritarianism, their commitment to intersectionality, avocation of alliance-politics, certainly find common ground in feminist and anarchist perspectives. The research that has emerged within this exciting field has been of considerable value in pushing knowledge and understanding about the animal condition, and how to respond to this, as well as building new and important bridges within academic and activist communities. This has included notable contributions focusing on queer theory and anti-speciesist praxis (e.g., Grubbs 2012); critical bioethics (e.g., Twine 2010); anthroparchy (Calvo 2008); the myth of sustainable meat (Stănescu 2010); and women of color and critical animal studies (Yarborough and Thomas 2010) that continue to promote understanding toward intersectionality, posthumanism, political economy, animal ethics, body commodification, alliance politics; and human, animal, and Earth liberation movements.

Certainly, there is a great deal that animal geographers can bring to CAS by foregrounding overriding questions of space and place to their analysis. To give one particularly important example, there is much work needed to further problematize the *geographies* of those deliberately hidden, out-of-sight and private spaces upon which terrifying amounts of violence, abuse, and killing of nonhuman animals – as well as serious forms of human exploitation – occur. As Carol Adams observed: "Indeed, patriarchal culture surrounds actual butchering with silence. Geographically, slaughterhouses are cloistered. We do not see or hear what transpires there" (1998: 49).

Perhaps one of the most important tasks for critical animal geographies is to continue to develop ways that meaningfully engage with those who, often at great personal cost to themselves, engage in direct action to transgress upon these private cloistered spaces of animal abuse. Ultimately, how can academics relate to animal activists in a way that allows them (beyond writing and research) to become better activists themselves? Again these are important ongoing questions which CAS fully embraces:

> CAS is unique in its defence of direct action tactics, its willingness to engage and debate controversial issues such as anti-capitalism, academic repressions, and the use of sabotage as a resistance tactic; its emphasis on the need for total liberation stressing the commonalities binding various oppressed groups; and the importance of learning from and with activists.
> (Best 2009: 13)

In this context, the final section of this chapter argues that more critical animal geography needs to engage in resistance, and, in doing so, can learn a lot from the nonviolent principles of anarchist activists. Drawing on anarchist and CAS theory moves to focus in more detail on (1) the importance of direct action, and the need for consistency between means and ends and; (2) the question of vegan praxis.

Critical animal geographies and direct action

It is vital that more critical animal geographies seek to further understand, explore, and engage (in every sense of the word) with the diverse activist geographies that underpin nonhuman animal and Earth liberation movements. In so many ways, *space* is absolutely central to understanding how (non)humans are currently abused, and how escaping the spaces in which they find themselves is deeply intertwined with their liberation. Contrasting ideas about captive spaces is also a fundamental division within the animal movement. For example, unlike those who identify with more welfarist types of reform for animals, CAS "rejects the exploitation of animal life as an ontological, epistemological and ethico-political strategy" (Jenkins and Stănescu 2014: 74), and agitates for the abolition of all these spaces of captivity and abuse: seeking empty cages, not larger cages (see Regan 2004).

Many activists and organizations aligned with the animal liberation movement identify with – or take inspiration from – anarchist praxis (see White and Cudworth 2014). Importantly, despite popular misrepresentations of anarchism as a synonym for violence and aggression, it is truer to say that anarchism encourages an ethics of nonviolence (e.g., see Baldelli 1971; Springer 2014). This call for nonviolence has a rich history in anarchist praxis. It is often overlooked that the Russian anarchist Lev Tolstoy's rejection of violence and thoughts toward civil disobedience were highly influential in informing the ideas of resistance adopted by leading figures such as Ghandi and Martin Luther King and their followers (see Jahanbegloo 2014). For anarchism and its commitment to prefigurative politics, it is vital that there is consistency between means and ends: violence and coercion are themselves aggressive forms of domination. Where violence is, anarchism is not.

Across the world, animal activists who have transgressed private, out-of-sight and hidden spaces of animal abuse – through multiple forms of direct (nonviolent, illegal) action – have done much to raise (public) consciousness about the plight of animals. Focusing on farmed animals, for example, high-profile undercover investigations have exposed and revealed sickening and horrific acts of cruelty against baby calves, cows, chickens, turkeys, pigs, and horses, including investigations by *Mercy for Animals* (www.mercyforanimals.org/investigations. aspx); Animal Rights Alliance (Sweden – www.ettlivsomgris.se/english); Hillside Animal Sanctuary (www.hillside.org.uk/); Vegetarians' International Voice for Animals (VIVA) (www.viva.org.uk/search/node/undercover); and Animal Aid (www.animalaid.org.uk/). All of these firsthand testimonies to extreme abuse and suffering in private farms and abattoirs provoked important media coverage, national outrage, and in several cases prosecution. Indeed Animal Aid, in an attempt to make visible these hidden spaces, is campaigning for mandatory CCTV in all UK slaughterhouses, which has met with some success. In the UK, "one in five abattoirs killing cows, pigs, goats and sheep now have CCTV" (Meikle: 2011, np).

These investigations have generated significant negative publicity and lost earnings. In response, powerful organizations involved in the animal abuse industry are making increasingly desperate attempts to further criminalize these forms of direct action. In addition to branding animal and environmentalist activists as terrorists, or ecoterrorists, a raft of punitive laws to prevent these accusing truths have been forthcoming. In the USA, for example, this included the Animal Enterprise Terrorism Act (AETA) in 2006, and more recently the so-called ag-gag laws (for a detailed critique of ag-gag laws, see Rasmussen, this volume). Essentially these laws are "designed to penalise investigative reporters who explore conditions on industrial agriculture operations.... It should come as no surprise to learn that the source of the pressure behind ag gag laws is, of course, industrial agriculture" (Smith 2012). With support for these laws being found in other countries, including Australia, the reality is that animal activists and the use of nonviolent acts of direct action are truly under siege (see Potter 2012). Against this background, it is imperative that more (animal) geographers engage with these matters seriously, and offer their voice and work in support and solidity where possible.

Critical animal geographies and vegan praxis

Before concluding I want to emphasize – as all CAS scholars do (Twine 2010) – the importance of an ethical vegan praxis. It is important to understand that critical veganism, and vegan praxis, embodies *far more* than just personal consumptive or lifestyle choices. As Harper (2010: 5–6) argues:

> Practitioners of veganism abstain from animal consumption (dietary and non-dietary). However, the culture of veganism itself is not a monolith and is composed of many different subcultures and philosophies throughout the world, ranging from punk strict vegans for animal rights, to people who are dietary vegans for personal health reasons, to people who practice veganism for religious and spiritual reasons.... Veganism is not just about the abstinence of animal consumption; it is about the ongoing struggle to produce socio-spatial epistemologies of consumption that lead to cultural and spatial change.

Writing before the word 'vegan' had been invented, Reclus (2013: 161) spoke of embracing "the principle of vegetarianism" as a response to the ugliness of animal abuse, and a means of becoming beautiful. For critical animal scholars, there is a constant need to pay attention to our own (personal) praxis, and recognize the need for consistency between means and ends: not contribute to those systems of domination that we write (or speak) against in our work. This is in many ways an important form of direct action, not least as it both embodies critical animal theory *and* confronts and transforms dominant human–animal binaries and speciesist hierarchies. For Torres (2007: 130) "Anarchist or not, anyone concerned about the cruelty animals experience at human hands should take the first and immediate step to stem that suffering by going vegan." Though, as always, the critical animal anarchist geographer would also be ever mindful of the complex commonalities of oppression.

Vegan praxis must incorporate a discourse and affect that reflect not only animal liberation but also total liberation. Vegan praxis must be orientated toward challenging all oppressive power structures.... A vegan praxis, ideally, is an ever-changing way of understanding and relating to oneself and all other beings based on empathy, authenticity, reciprocity, justice and integrity – the principles that underscore true freedom.

(Weitzenfeld and Joy 2014: 25)

Similarly, in a later addendum to Dominick's (1997: 21) highly influential paper on veganism and anarchism he notes:

I'm the first to be disgusted by those stodgy radicals, mostly of the 'old school,' who proclaim lifestyle changes must, at the very least, take a back seat to the 'real' work of social change, which is limited to the restructuring of social institutions. Still, their critique of those who, on the opposite end, believe personal change will actually be the revolution when practiced on a large scale, is rather important. We must avoid either extreme.... [T]he simple act of changing one's lifestyle, even when joined by millions of others, cannot change the world, the social structures of which were hand-crafted by elites to serve their own interests.

Where possible, it is this richer, more radical and integrated appreciation of vegan praxis – which is far more than consumption choices – that a critical animal geography sensibility should encourage and work toward. In so many ways this approach has the potential to help usher in new and important counter-hegemonic spaces of resistance: spaces of resistance created through embracing a politics of peace and nonviolence. Two excellent examples that illustrate the tensions between theory and practice in food-activism movements, namely the transnational Food Not Bombs network and UK-based Nottingham Vegan Campaigns, are discussed in the following chapter by Giraud (this volume).

Critical animal geography, CAS and anarchist praxis: some final thoughts

We must take sides. Neutrality helps the oppressor, never the victim. Silence encourages the tormentor, never the tormented.... Sometimes we must interfere.

(Elie Wiesel (1986))

The call for new critical sites of resistance and counter-hegemonic spaces with which to address the animal plight in contemporary society is desperately urgent. To these ends this chapter has sought to encourage more animal geographers to seek inspiration from, and embed their future research clearly within, anarchist praxis and the related field of Critical Animal Studies. Being inspired by anarchist praxis in particular, and responding to its unique commitment to challenging multiple sites of oppression, placing great emphasis on intersection-ality, nonviolent strategies of resistance, and a vegan praxis, promises to usher in exciting new synergies not only within the various branches of geography, but between other academic disciplines, as well as activist and wider public circles. Of course, there are significant gaps in the anarchist and CAS literature that should be addressed, not least those that address the domination and exploitation within and between species as a means of further problematizing human power and human species identity. Recognizing the commonality of oppression, a pol-itics of total liberation – of humans, nonhuman animals, and the Earth – should be kept in mind at all times. A critical focus on the nonhuman animal question must not be seen as either indulgent, or of secondary interest to the many important 'human problems' that geographers are engaged with. As Ryder (2000: 1) argued: "The struggle against speciesism is not a side-show; it is one of the main arenas of moral and psychological change in the world today. It is part of a new and enlarged vision of peace and happiness."

It is one thing to write about critical animal geography inspired by anarchist lines of flight; it is another thing to put this into practice. Encouraging new net-works of solidarity and support to gain traction may be frustratingly slow, even in the *critical* spaces where we could rightly expect it to be strongest. Certainly, constructing necessary bridges across academic, activist, and wider communities to help further liberate (other) animals from the human hells they suffer will be built on a remarkable level of commitment, patience, tolerance, and respect. For the future of animal geography, my hope is that Reclus's transformative call to become beautiful ourselves, both in our relations with each other and with other animals, comes to define the very heart and spirit of a *critical* animal geography. The belief and expectation that this future will be possible, enactable and achiev-able prevails.

Acknowledgments

I would like to thank Kathyrn Gillespie and Rosemary Collard for their excellent guidance, feedback, and encouragement during the writing of this chapter.

References

Adams, C 1998, *The sexual politics of meat: A feminist-vegetarian critical theory*, Con-tinuum International Publishing Group, New York.
Baldelli, G 1971, *Social anarchism*, Penguin Books, Middlesex.
Best, S 2009, 'The rise of critical animal studies: Putting theory into action and animal liberation into higher education,' *Journal for Critical Animal Studies*, vol. VII (1), pp. 39–52.
Best, S, Nocella, AJ, Kahn, R, Gigliotti, C, and Kemmerer, L 2007, 'Introducing critical animal studies,' *Journal for Critical Animal Studies*, vol. 1 (1), pp. 4–5. Viewed January 22, 2014. www.criticalanimalstudies.org/wp-content/uploads/2009/09/Introducing-Critical-Animal-Studies-2007.pdf

Bookchin, M 1993, 'Deep Ecology, anarchosyndicalism and the future of anarchist thought,' in M Bookchin, G Purchase, B Morris, and R Aitchtey (eds), *Deep ecology & anarchism*, Freedom Press, London.

Braun, B 2005, 'Environmental issues: Writing a more-than-human urban geography,' *Progress in Human Geography*, vol. 29 (5), pp. 635–650.

Buller, H 2013a, 'Animal geographies I,' *Progress in Human Geography*, vol. 38 (2), pp. 308–318.

Buller, H 2013b, 'Individuation, the mass and farm animals,' *Theory Culture & Society*, vol. 30 (7/8), pp. 155–175.

Calvo, E 2008, '"Most farmers prefer Blondes": The dynamics of anthroparchy in animals' becoming meat,' *Journal for Critical Animal Studies*, vol. VI (1), pp. 32–45.

Dominick, B 1997, *Animal Liberation and Social Revolution*. Viewed January 24, 2014. http://theanarchistlibrary.org/library/brian-a-dominick-animal-liberation-and-social-revolution.

Emel, J, Wilbert, C, and Wolch, J 2002, 'Animal geographies,' *Society and Animals*, vol. 10 (4), pp. 407–412.

Gillespie, K 2013, 'Sexualized violence and the gendered commodification of the animal body in Pacific Northwest, US dairy production,' *Gender, Place and Culture*. Viewed January 31, 2014. DOI: 10.1080/0966369X.2013.832665.

Giraud, E 2015, 'Practice as Theory: learning from food activism and performative protest,' in K Gillespie and R-C Collard (eds), *Critical Animal Geographies*, Routledge, Abingdon.

Goldman, E 2005, 'Voltairine De Cleyre,' in S Presley and C Sartwell (eds), *Exquisite Rebel: the essays of Voltairine de Cleyre – Anarchist, Feminist, Genius*, New York Press, Albany.

Grubbs, J 2012, 'Guest Editorial: Queering the que(e)ry of Speciesism,' *Journal for Critical Animal Studies*, vol. 10 (3), pp. 4–6.

Harper, AB 2010, 'Race as a "feeble matter" in Veganism: Interrogating whiteness, geopolitical privilege, and consumption philosophy of "cruelty-free" products,' *Journal for Critical Animal Studies*, vol. III (3), pp. 5–27.

Herzog, HH 2007, 'Gender difference in human–animal interactions: a review,' *Anthrozoos*, vol. 20 (1), pp. 7–21.

Hobson, K 2007, 'Political animals? On animals as subjects in an enlarged human geography,' *Political Geography*, vol. 26 (3), pp. 250–267.

Jahanbegloo, R 2014, *Introduction to nonviolence*, Palgrave Macmillan, Basingstoke.

Jenkins, S and Stănescu, V 2014, 'One Struggle,' in A Nocella, J Sorenson, K Socha, and A Matsuika (eds), *Critical animal studies reader: An Introduction to an intersectional social justice approach to animal liberation*, Peter Lang Publishing Group, New York.

Johnston, C 2008, 'Beyond the clearing: Towards a dwelt animal geography,' *Progress in Human Geography*, vol. 32 (5), pp. 633–649.

Jones, O 2000, 'Inhuman geographies: (un)ethical spaces of human non-human relations,' in C Philo and C Wilbert (eds), *Animal spaces, beastly places: New geographies of human-animal relations*, pp. 268–291, Routledge, London.

Kropotkin, P 1988 [1902], *Mutual aid: A factor of evolution*, Freedom Press, London.

Masson, JM 2004, 'Foreword' in T Regan, *Empty Cages: facing the challenge of animal rights*, Rowman & Littlefield, Oxford.

Matless, D 1994, 'Moral geography in Broadland,' *Ecumene*, vol. 1, pp. 127–156.

Meikle, J 2011, 'Slaughterhouses could be forced to fit CCTV to prevent animal abuse,'

Guardian. Viewed June 23, 2014. www.theguardian.com/world/2011/nov/08/slaughterhouses-cctv-prevent-abuse.

Mitchell, L 2011, 'Moral disengagement and support for nonhuman animal farming,' *Society & Animals*, vol. 19, pp. 38–58. Viewed January 30, 2014. www.animalsandsociety.org/assets/462_mitchellsa.pdf.

Nast, HJ 2006, 'Critical Pet Studies?,' *Antipode*, vol. 38 (5), pp. 894–906.

Nocella, AJ, Sorenson, J, Socha, K, and Matsuoka, A 2014, 'The emergence of critical animal studies: The rise of intersectional animal liberation,' in AJ Nocella II, J Sorenson, K Socha, and A Matsuika (eds), *Defining Critical Animal Studies: An intersectional social justice approach for liberation*, Peter Lang, New York.

Pedersen, H and Stănescu, V, 2012, 'What is "critical" about animal studies? From the animal "question" to the animal "condition",' in K Socha, *Women, destruction, and the avant-garde: A Paradigm for animal liberation*, pp. ix–xii, Rodopi Press, New York.

Peet, R 1975, 'For Kropotkin,' *Antipode*, vol. 7 (2), pp. 42–43.

Potter, W 2011, *Green is the new red: an insider's account of a social movement under siege*, City Lights Books, San Francisco.

Rasmussen, C 2015, 'Pleasure, Pain and Place: Ag-gag, crush videos, and animal bodies on display,' in K Gillespie and R-C Collard (eds), *Critical Animal Geographies*, Routledge, Abingdon.

Reclus, E 2013 [1901], 'On Vegetarianism,' in J Clark and C Martin (eds), *Anarchy, geography, modernity: selected writings of Elisée Reclus*, PM Press, Oakland.

Regan, T 2001, *Defending Animal Rights*, University of Illinois Press, Illinois.

Regan, T 2004, *Empty Cages: facing the challenge of animal rights*, Rowman & Littlefield Publishers, Oxford.

Roe, E 2010, 'Ethics and the non-human: The matterings of sentience in the meat industry,' in B Anderson and P Harrison (eds), *Taking-place: Non-representational theories and geographies*, pp. 261–280, Farnham, Ashgate.

Ryder, RD 1998, 'Painism,' in M Bekoff and CA Meaney (eds), *Encyclopedia of animal rights and animal welfare*, pp. 269–270, Greenwood Press, Westport.

Ryder, RD 2000, *Animal revolution: Changing attitudes towards speciesism*, Berg, Oxford.

Ryder, RD 2010, *Speciesism, painism and happiness: A morality for the twenty-first century*, Imprint Academic, Exeter, UK and Charlottesville, VA.

Scully, M 2002, *Dominion: The power of man, the suffering of animals, and the call to mercy*, St Martin's Press, New York.

Singer, P 1995, *Animal Liberation 2nd Edition*, Pimlico, London.

Smith, SE 2012, 'Agriculture gag laws are violating press freedom in the US,' *Guardian*. Viewed July 6, 2014. www.theguardian.com/commentisfree/2012/jun/06/agriculture-gag-laws-press-freedom.

Socha, K 2012, *Women, destruction, and the avant-garde: A paradigm for animal liberation*, Rodopi, Amsterdam.

Springer, S 2013, 'Anarchism and Geography: A brief genealogy of anarchist geographies,' *Geography Compass*, vol 7 (1), pp. 46–60.

Springer, S 2014, 'War and pieces,' *Space and Polity*, vol. 18 (1), pp. 85–96.

Springer, S, Ince, A, Pickerill, J, Brown, G, and Barker, AJ 2012, 'Reanimating anarchist geographies: A new burst of colour,' *Antipode*, vol. 44 (5), pp. 1591–1604.

Stănescu, V 2010, '"Green" eggs and ham? The myth of sustainable meat and the danger of the local,' *Journal for Critical Animal Studies*, vol. VIII (1/2), pp. 8–33.

Taylor, N and Twine, R 2014, 'Locating the "critical" in critical animal studies,' in N Taylor and R Twine (eds), *The rise of critical animal studies: From the margins to the centre*, Routledge, Abingdon.

Tipper, B 2011, ' "A dog who I know quite well": everyday relationships between children and animals,' *Children's Geographies*, vol. 9 (2), pp. 145–165. DOI: 10.1080/14733285.2011.562378.

Torres, B 2007, *The political economy of animal rights*, AK Press, Edinburgh.

Twine, R 2010, *Animals as biotechnology: Ethics, sustainability and critical animal studies*, Routledge, New York.

Urbanik, J and Morgan, M 2013, 'A tale of tails: The place of dog parks in the urban imaginary,' *Geoforum*, vol. 44, pp. 292–302.

Ward, C 2004, *Anarchism: a very short introduction*, Oxford University Press, Oxford.

Weitzenfeld, A and Joy, M 2014, 'An overview of anthropocentrism, humanism, and speciesism in critical animal studies,' in A Nocella, J Sorenson, K Socha and A Matsuika (eds), *Critical animal studies reader: An introduction to an intersectional social justice approach to animal liberation*, Peter Lang Publishing Group, New York.

Whatmore, S and Thorne, LB 1998, 'Wild(er)ness: Reconfiguring the geographies of wildlife,' *Transactions of the Institute of British Geographer*, vol. 23 (4), pp. 435–454.

White, RJ and Cudworth, E 2014, 'A challenge to systems of domination: From corporations to capitalism,' in A Nocella, J Sorenson, K Socha, and A Matsuika (eds), *Critical Animal studies reader: An introduction to an intersectional social justice approach to animal liberation*, Peter Lang Publishing Group, New York.

Weick, D 1971, 'Preface,' in G Baldelli, *Social anarchism*, Penguin, Harmondsworth.

Wiesel, E 1986, Nobel Peace Prize Acceptance Speech, December 10, Oslo City Hall, Norway.

Wolch, J 2002, 'Anima Urbis,' *Progress in Human Geography*, vol. 26 (6), pp. 721–742.

Wolch, J and Emil, J (eds) 1998, *Animal geographies: Place, politics and identity in the nature-culture borderlands*, Verso, London.

Yarborough, A and Thomas, S (2010), 'Women of color in critical animal studies,' *Journal for Critical Animal Studies*, vol. VIII (3), pp. 3–4.

3 Practice as theory

Learning from food activism and performative protest

Eva Giraud

This chapter draws on two food-activism movements, the transnational Food Not Bombs network and UK-based Nottingham Vegan Campaigns, to examine how activists' navigation of tensions between theory and practice can inform a critical animal geography.

I suggest that engaging with activists – and particularly participating in projects where complex political issues are articulated in practice – provides valuable insights about how to politicize theories that have proven popular in 'third-wave' animal geography, such as posthumanism and nonrepresentational theory (Urbanik 2013; Buller 2014). I argue, more specifically, that taking the work of activists seriously as theory is useful in balancing the need for concrete action demanded by critical animal studies (CAS) (Best 2009; Weisberg 2009; Dell'Aversano 2010), and the decentering of the human called for by animal geographies and 'mainstream' animal studies more broadly (Wilson *et al.* 2011; Lorimer 2013). A focus on activism is thus useful in generating dialogue between theoretical perspectives that are often opposed to one another and, in so doing, can help to inform a more critical animal geography, due to the insights these perspectives can bring to one another when placed in conversation.

Food Not Bombs has already been written about extensively elsewhere (Mitchell and Heynen 2009; Heynen 2010; Sbicca 2013), as well as within self-reflexive analyses from activists involved (Crass 1995; McHenry 2012), so will be used here as a contextual framework for discussing the tradition that shapes Nottingham Vegan Campaigns. Due to my involvement with the UK-based movement, for the second case study I will draw on auto-ethnographic experiences (taking a lead from Mason 2013) to reflect on tensions between political ideals and logistics within these campaigns and how these protests can be seen as negotiating the action demanded by CAS with the challenge to human exceptionalism posed by animal geographies. I reflect in particular on my own experiences of developing tactics to combat power relations between activists and other parties involved in the events; of using food to open space for dialogue with diverse publics; and of the affective environment generated by the protests.

Critical animal studies and posthumanism

Before exploring *how* activism can create dialogue between critical animal studies (CAS) and posthumanism, it is necessary to ask *why* it is valuable to create this dialogue. Though animal activism cannot be treated as synonymous with CAS as an academic field, they share important affinities. What characterizes CAS is its central concern with contesting existing human–animal relations – which are framed as exploitative – and its demand for concrete action (Best 2009, see White this volume). The field, moreover, emerged from a 'radical animal liberation' tradition and maintains a firm link between theory and praxis, coupled with a commitment to an 'abolitionist perspective' that rejects practices seen to benefit humans at the expense of animals (Pedersen 2011: 66–67).

CAS has been criticized, however, for having a totalizing stance towards human engagements with animals, for positioning 'the animal' as a pure category that should not be interfered with by humans, and for denying the complexity of human–animal relations (Haraway 2008: 299). In focusing on animals CAS has also been accused of reinforcing the same exceptionalist logic that underpins human privilege, by grounding "its appeals for animal rights on the comparable existence of essential human characteristics ... in non-humans – extending the franchise to certain privileged others" (Lorimer 2013: 12). While, as touched on below, these arguments elide the more nuanced arguments made by CAS,[1] they nonetheless foreground how power relations within existing political frameworks can lead to inadvertent anthropocentrism.

Posthumanism, in contrast, actively challenges human exceptionalism but has been accused of depoliticizing animal studies, through displacing political frameworks that could contest exploitation (Adams 2006; Weisberg 2009). Posthumanist approaches have productively unsettled human privilege by focusing on how the social emerges through complex networks of interaction between human and nonhuman entities (inspired by thinkers such as Latour 2005). 'The human,' moreover, can only exist through its relation with other actors, in line with Barad's (2007) concept of intra-action (or the notion that entities do not ontologically preexist one another, but emerge as distinct entities through their relations). These theories have proven useful in informing animal geography, by not only challenging the ontological status of the human as separate from 'nature,' but – as Wilson (2009) argues – posing an important epistemological challenge to knowledge frameworks that privilege the human.

From this perspective, arguments such as those criticized by Lynn – which frame animals as "resources that lay beyond the boundaries of moral community" (1998: 281) – are untenable; 'the human' cannot be separated out as a distinct category, worthy of special treatment, in order to justify exploitative relationships with other entities. Instead, "cosmopolitical" approaches to politics are advocated (Stengers 2010, 2011), which don't begin with fixed categories (such as human and animal) that predetermine who has agency, privilege or the right to speak, and instead take the 'risk' of experimenting with new political configurations (Bird Rose 2012; Buller 2014). Related theoretical work such as nonrepresentational theory (Thrift 2008; Anderson and Harrison 2010) similarly stresses the agency of all nonhuman entities – from stones to insects (Bennett 2010) – and highlights the difficulties in speaking for, or representing, the non-human without installing hierarchical relationships that perpetuate human privilege (Roe 2010).

These theories thus provide insight into the co-constitutive processes through which animals are categorized as commodities – and humans as consumers of these commodities – within the agricultural-industrial complex. Understanding these identities as relational, rather than essential, actively disrupts the inevitability of these categorizations and opens space for new forms of political engagement. In making this argument, however, these theories also problematize more radical political frameworks that distinguish between different categories of actors, in order to defend the rights of these actors. While human exceptionalism is challenged, therefore, animals are also prevented from being set apart as a special category worthy of protection, and – from a CAS perspective – this has precluded a more practical understanding of how to intervene in exploitative practices.

Although the "relational turn" in geography has proven valuable for unsettling human privilege (Buller 2014: 314), therefore, it has also been treated cautiously by geographers who are attuned to the power relations embedded within these relationships. The problem, when 'relating' is uncritically celebrated, is that this fails to guard against colonizing relationships between 'humans' and 'nature,' especially when these concepts themselves are destabilized (Lorimer 2013: 11). For this reason geographers have argued it is vital that: "The ethical potential that proximity and conjoining bring must be held in productive tension with acknowledgements of animals' spatial and subjective requirements" (Collard 2014: 162). Whilst avoiding essentialist characterizations of nonhuman life, therefore, the *critical* use of categories that stress the "autonomy and alterity" of nonhumans can still be useful in guarding against exploitation, as illustrated by Collard's recuperation of the concept of 'wildness': "The point is not to imply that wildlife can only exist 'out there,' away from humans, but rather that it might require a degree of freedom that controlled (or even forced) proximity with humans does not permit" (Collard 2014: 114).

To take these arguments further, the active contestation of animal exploitation (provided by CAS) should also not be dismissed lightly. Even though distinguishing animals as a special category might reinforce essentialist values, the problem is that this categorization has already occurred via our intra-actions with animals within the agricultural-industrial complex, where they emerge as commodities (Ufkes 1998; Shukin 2009). Although posthumanism is useful in denaturalizing hierarchical relations and highlighting *how* animal-commodity and human-consumer categories are actively produced through relationships forged within the agricultural-industrial complex, these processes also require material disruption. In other words, as these categories are created through intra-action, existing human–animal relations require active contestation in order to unsettle the epistemological positioning of animals as 'exploitable.'

Hints for how to reconcile CAS's demands for concrete action with posthumanism's decentering of the human, without perpetuating essentialist notions of the animal, can be found in Buller's argument that the potential for a radical politics to emerge from animal geography lies in: "the political expression and mobilization of this emergent relational ontology" (2014: 314). In other words, the task is to find a means of politicizing approaches that stress relationality and putting them to practical work. It is in addressing this task that dialogue between posthumanism and CAS is particularly useful, and focusing on activism is a valuable means of opening this conversation.

Posthuman politics and activist praxis

Though often loosely referred to as 'anarchist' (Crass 1995), a more precise characterization of the food-activist groups at stake here would be as autonomous social movements (Pickerill and Chatterton 2006). What is significant about this understanding of 'autonomy' is that it dovetails with debates in animal studies. Traditions of autonomy within activism are distinct from the term's use in reference to liberal-individualism (where people are seen as possessive individuals with the 'right' to do what they want) instead referring to collective practices that are autonomous from capitalism (Chatterton and Pickerill 2010), stressing the need for responsibility between members of these communities and necessity of continuous reflection on how individual actions impact upon others. The difficulty is in realizing these principles in practice, as this involves crafting alternatives that do not draw on preexisting political frameworks (such as representational democracy, party or vanguardist politics), as these structures are seen as grounded in liberal-humanist traditions that perpetuate the privilege of certain social groups (Juris 2005; Nunes 2005).

Whilst posthumanist theory and autonomous praxis might seem like unlikely bedfellows, in light of tensions between radical geography and nonrepresentational theory (Somdahl-Sands 2013), what is being grappled with by both is the need to challenge both normative conceptions of the subject as the locus for politics and normative political and ethical values (that result in social and epistemological hierarchies). Indeed these shared affinities have caused new social movements to be a source of inspiration for Stengers (2011) and (to a lesser extent) Haraway (2008: 3). For autonomous activism the difficulty is in developing alternatives to capitalism that do not reproduce the social and cultural hierarchies (such as problematic gender, race, and class relations) of the system being challenged, but still maintain space for concrete action. Similarly, the problem for posthumanism is in developing a politics to confront contexts in which an anthropocentric ethics is no longer viable (where liberal-humanism and related concepts such as 'rights' have been problematized), but concrete action is still needed (Braidotti 2013).

The difficulties in crafting such a politics are brought into focus when considering the task faced by animal rights activists, who are confronted with specific problems in realizing nonrepresentational politics. Within autonomous praxis, certain mechanisms (such as consensus decision-making) have been developed to maximize participation by creating space for marginalized individuals to have a voice rather than just trying to represent their interests, even if these processes are complicated to implement (Robinson and Tormey 2007). This is more problematic to realize when the parties that need to be involved are not human, so demands cannot be dialogically formulated in the manner that has been integral to direct-democracy.

Although a growing body of research is exploring nonverbal means of engaging with animals in order to coproduce knowledge (Greenhough and Roe 2011; Despret 2013; Latimer and Miele 2013), as pointed out in recent debates about the value of Haraway to cultural geography (Roe 2010; Wilson *et al.* 2011), work still needs to be done in exploring how to accomplish this when direct engagement is impossible. This problem is compounded in situations (such as the U.S. and UK urban contexts in which the activists examined here are working), where certain types of animals have been systematically rendered invisible as the spaces in which livestock are sold and slaughtered have been removed from the city throughout the later nineteenth and early twentieth centuries (Philo 1995). Both Food Not Bombs and Nottingham Vegan Campaigns, therefore, are faced with the difficult task of making these relations visible, without falling back onto representational political frameworks.

Food Not Bombs

Food Not Bombs is a transnational movement that shares free vegetarian food in city center locations. The food produced by Food Not Bombs is 'gleaned' from industrial bins filled with discarded products from supermarkets and these edible foods are then prepared and shared with local homeless populations (Sbicca 2013). These events synthesize food-sharing with performative protest, drawing together a series of political issues relating to corporate waste, the economic and social exclusion of certain populations from the city, food poverty, and the excesses of the agricultural-industrial complex. The significance of food-sharing is – in Clark's words – because "as a site of resource allocation, food tends to recapitulate power relations" (2004: 22). For Food Not Bombs sharing food that supermarkets treat as 'waste' with populations who suffer food poverty is designed to highlight the contradictions of consumer capitalism, where excess food is produced but people are still left hungry (Sbicca 2013). By sharing food in city center locations, these contradictions can be highlighted in commercial spaces where they are ordinarily naturalized. The protests, therefore, use food as a lens through which to articulate how a range of sociopolitical issues coalesce around questions of food production and consumption.

In U.S. contexts food-sharing foregrounds a series of social exclusions engendered by neoliberal economic policy, which has removed "social safety nets and programs that reduce poverty" (Sbicca 2013: 4) and led to legislation that discourages homeless populations from inhabiting public space (Mitchell and Heynen 2009: 620). These trends reinforce commercial rhythms by removing people whose presence could denaturalize the everyday practices of consumer capitalism (Heynen 2010: 1229–1230). Such policies are what Food Not Bombs contest, with the protests foregrounding food poverty whilst simultaneously inviting marginalized populations to reclaim public space and "participate in the work and the making of the city and the right to urban life (which is to say the right to be part of the city – to be present, to *be*)" (Mitchell and Heynen 2009: 616, italics in original). The role of the protests is thus to make issues that are usually hidden visible (Mitchell and Heynen 2009; Heynen 2010).

It is not just the exclusion of homeless populations that is made visible, however, as the food is created from ingredients that have been expelled from commercial spaces, in order to highlight waste. Thus, even though the food is vegetarian or vegan, gleaning is in explicit resistance to the "yuppie health food" conceptions of veganism that have been recuperated as ethical lifestylism (Clark 2004: 26). In line with DiVito Wilson (2013), instead of describing groups such as Food Not Bombs as "alternative food networks" it is therefore important to position them as radical-autonomous projects, distinguishing them from highly classed food practices that have often been cited as alternatives to 'Global Food' (the examples Carolan (2011) gives being heritage seed banks, organic farming cooperatives and backyard chicken coops).

These tactics, however, also illustrate dangers associated with the desire to represent particular issues or populations, as foregrounded by both posthumanism and nonrepresentational theory as well as by autonomous praxis. For example, Haraway (1992, 2011) argues that the desire to represent can lead to inadvertent political ventriloquism, and Nunes (2005) raises the danger of informal hierarchies that has been a preoccupation of activists. In the case of Food Not Bombs, this has sometimes manifested itself as hierarchical relations between activists and homeless people; when food-sharing was banned in Orlando, for instance, activists were keen to maintain the protests and face arrests, whereas more vulnerable homeless populations were not able to take this risk (Sbicca 2013: 9). There is also the broader danger of food-sharing becoming a political statement, which serves an anticapitalist activist agenda, and interpellates hungry people into activist performances (Heynen 2010: 1228). The protests thus run the constant danger of slipping back into representational modes of politics, which reinforce rather than challenge social inequalities.

In line with Chatterton and Pickerill's emphasis on the reflexive nature of autonomous politics (2010), however, activists who acknowledge these dangers deploy several tactics to attenuate these problems. First, they invite people they are sharing food with to participate in every aspect of the protests (from gleaning to cooking) (Mitchell and Heynen 2009). Second, they explicitly frame their actions as 'sharing' food rather than 'distributing' it, in an attempt to challenge inequalities between activists and those consuming the food (Heynen 2010: 1228). Activists therefore try to make the protests participatory and create space for the other parties involved to actively make a difference in shaping public space (rather than simply following the activists' lead). In Chatterton and Pickerill's terms this is invariably a "messy" process, but in contesting commercial rhythms by coproducing "geographies of explicitly mutual aid" (Mitchell and Heynen 2009: 626), the protests can be seen as cosmopolitics in action. In publicly and visibly sharing food, activists open themselves up to the 'risk' of having their actions and preconceptions shaped by the diverse publics who engage in the protests with them.

Despite being characterized by some of the tensions faced by autonomous activism more broadly, therefore, Food Not Bombs still provides valuable insights for moving beyond representational politics, whilst still maintaining a commitment to concrete, interventional action. The protests explicitly intervene in the everyday practices that occur within urban space by using food as the nexus for intersecting forms of exploitation. They do this by constructing a new geography of urban space that is premised on an alternative logic, which directly contests the commercial forces that shape these sites and reintroduces actors that have systematically been rendered invisible by these forces (from homeless populations to discarded food). Crucially, though, these performances do not rely on activists representing these issues, but coproduce alternative food geographies with actors who are ordinarily excluded from these spaces.

For the purpose of animal geography, further work needs to be done to elucidate the role of veganism and vegetarianism within food activism. Resonating with the centrality of veganism within CAS, where it is seen as a concrete means of challenging exploitation, the rejection of animal products is one of the three principles of Food Not Bombs (along with consensus decision-making and nonviolence; Crass 1995: 4). The role of food requires further exploration, however, in light of criticisms levelled at vegan praxis both within activism and posthumanist theory.

Vegan praxis

Research on ethical vegetarianism has foregrounded its value as a disruptive practice, which departs from normative food-consumption practices in the global north (Dietz *et al.* 1995; Fox and Ward 2008; Kwan and Roth 2011). Veganism in particular has been associated with radical politics, as a marker of anarchist identity (Clark 2004; Portwood-Stacer 2012). These findings resonate with activists' own positioning of veganism as a situated response to the animal exploitation that intersects with other exploitative social relations (Giraud 2013b). In activist reflection on Food Not Bombs, for instance, the role of veganism is described as "a political act against the meat and dairy industries and to promote ecological sustainability, equal distribution of food and resources throughout the world, human health and animal liberation" (Crass 1995: 6).

These arguments, however, overlook the contentious role of veganism within activist praxis. Clark, for example, points out that veganism "reveals ideological fissures within [anarcho-] punk culture," due to its normative status, leading to individuals "flaunt[ing] meat-eating as a way of challenging punk orthodoxy" (2004: 23). As I have outlined elsewhere, it is also the locus for debates within autonomous activism where it is seen as something that has become unquestionable, with the potential to inadvertently exclude certain social groups who are unfamiliar both with the practice and the political rationale underpinning it (Giraud 2013b). Such concerns intersect with framings of veganism that appear within 'mainstream' animal studies, such as Haraway's concerns about it becoming a totalizing norm that forecloses context-specific relations with animals (2008: 80), or in cultural geography, such as Guthman's suggestion that it is highly classed (2008).

Paying closer attention to how veganism is actually articulated in food protests provides an alternative perspective, however, as in these contexts it is enacted less as a totalizing imperative and more as what Kheel characterizes as an "invitational approach" to engagement with animal rights (2004: 335). By sharing food with the public, people are invited to engage in practices that are antithetical to those promoted by the agricultural-industrial complex and to engage in dialogue with activists about these foods. These practices, therefore, could be understood as a form of prefigurative politics – that uses concrete action as a basis for exploring alternative ways of living – rather than the imposition of activist norms and values.

Nottingham Vegan Campaigns

Nottingham Vegan Campaigns emerged after activists, formerly involved with biannual protests against McDonald's, began to incorporate food giveaways into their protests. These events were held on an irregular basis from spring 2008 onwards, but throughout 2010 we organized monthly protests that culminated in a multi-target protest, which consisted of five food stalls and five activist literature stalls that were spread across the city (Veggies 2014a, see Figure 3.1). A focus on these protests is useful in indicating how practices advocated by CAS (such as veganism) could complement the decentering of the human that has been central to recent work in animal geography. Or, from a different perspective, these protests offer insights from activist praxis which could politicize posthumanism and related theories and (to go back to Buller's argument) find "political expression" for an "emergent relational ontology" (2014: 314). Before exploring how these protests can inform theoretical work, however, it is useful to establish their context.

Food giveaways are a central part of grassroots vegan advocacy practices in the UK (Vegan Society 2014, Veggies 2014b), with numerous local groups and the Vegan Society themselves supplying resources to aid local initiatives (indeed we received a small grant from the society to fund Nottingham events). This background of outreach, however, could be seen as problematic in undermining some of the key tenets of autonomous praxis, as set out by Nunes's argument that "nothing is what democracy looks like.... By deciding on an ideal model of what it should be like, all we are doing is creating a transcendent image that hovers above actual practices" (2005: 310–311). If veganism becomes what a politically radical diet "looks like," therefore, it is in danger of being incompatible with autonomous ideals, as well as being the sort of 'totalizing' value that is criticized by posthumanism.

Figure 3.1 Nottingham Vegan Campaigns' food giveaway outside McDonald's (image ©
Eva Giraud).

The Campaigns' associations with outreach, however, should not overshadow
the influence of autonomous activism on the protests. Nottingham Vegan Campaigns was initiated by members of long-standing Nottingham catering collective, Veggies, whose campaigning history informed the protests' tactics and
scope. Since their launch in 1984 Veggies have had a long history of involvement in grassroots protest (with *Peace News* describing them as "the field

kitchen of the UK activist movement," Smith 2009). Most significantly, Veggies had a key role in UK anti-McDonald's campaigning during the late 1980s/early 1990s, which culminated in the 'McLibel' trial (1994–1997) where two activists were sued for their role in distributing a 'fact sheet' critical of the corporation. Veggies were also threatened with legal action for distributing their own pamphlet, but slightly altered its wording and continue to distribute it today (Vidal 1997; Wolfson 1999; Giraud 2008).

Veggies' *What's Wrong With McDonald's* pamphlet (n.d.) used McDonald's to argue that global food corporations foster exploitative social relations that implicate food workers, animals, and consumers. They argue that these corporations create new food geographies by transforming urban space on a local level (competing with local business, deskilling and depressing wages in the catering industry, generating increased levels of waste) and impacting on sites in the majority world that supply its products (from workers producing Happy Meal toys, to laborers working in ranches on ex-rainforest land).[2] From the outset Veggies sought to make these political arguments more accessible; indeed the collective originated from a playful symbolic critique of McDonald's, whereupon members "from Nottingham's animal rights group had the idea of presenting the manager of a local McDonalds with a huge veggie burger to represent an ethical alternative" (Veggies 2014c). The collective then established its own veggie burger van near McDonald's (1985–2000), to act as a direct counterpoint to McDonald's food (Smith 2011). In 2008 Nottingham Vegan Campaigns revived this idea by incorporating food giveaways into existing 'days of action' against McDonald's that were led by local animal rights groups. It was at this point that I became involved in the actions, going on to cofacilitate all of the 2010 protests. Prior to 2008 the anti-McDonald's days mainly consisted of distributing pamphlets, so we originally saw cooking veggie burgers and distributing soya shakes as a playful way of attracting more attention. The success of these tactics in facilitating dialogue with the public, however, led to us organizing giveaways on a more regular basis.

On one level the food giveaways dovetailed with the interventional tactics of Food Not Bombs, as sharing food in city center locations was a response to exclusions occurring at a local level in Nottingham. The multi-target protest of December 2010, for instance, was a specific challenge to Nottingham council's decision to prevent charitable or community groups using the Council House (a large, central, local government venue in Nottingham) and the prohibitive charges levied at using the city's market square for more formal events (although commercial organizations were still able to use both).[3] Like Food Not Bombs, therefore, Nottingham Vegan Campaigns sought to challenge the exclusion of local people from public space by sharing food outside the Council House and in the centre of the market square, amidst a commercial Christmas market (see Figure 3.2).

As with Food Not Bombs, however, certain logistical problems led to these protests having an uneasy relation with the systems they were contesting. While Food Not Bombs have been criticized for relying on waste from the system they

Figure 3.2 Nottingham Vegan Campaigns' food giveaway outside the Council House
(image © Eva Giraud).

are condemning (Sbicca 2013: 9), Nottingham Vegan Campaigns could be seen
as strengthening neoliberal logic. Instead of gleaning, the Nottingham protests
used food donated from local independent businesses and cooperatives, running
the danger of promoting localism and/or veganism as an 'ethical' alternative.
This approach could, therefore, be seen as problematically elitist and even as
engendering hierarchical relations between activists and publics under the guise

of "bringing good food to others" (Guthman 2008: 433). In the UK this was a particular problem due to similarities with political rhetoric by the right-wing Conservative government, whose 'Big Society' agenda (also launched in 2010) advocated rolling back the state and placing social responsibility on individuals, resonating with broader concerns about radical community initiatives compensating for neoliberal policy (Caffentzis and Federici 2014).

We developed several tactics to counter these problems. One of the central initiatives was facilitating skill-share workshops, and inviting members of the public to join us in preparing food. Even this, however, ran the danger of didacticism in light of highly classed cultural discourse surrounding healthy eating in the UK (Warin 2011; Piper 2013). Like Food Not Bombs we framed these workshops as sharing skills (rather than teaching them) to counter this danger, and ran the workshops for free in autonomous spaces. Even this approach had its drawbacks though, in making it difficult to involve nonactivist audiences. Whilst the protests themselves generated engagement with members of the public, few people were willing to actually enter activist space and participate in the skill-shares themselves. Throughout 2010, therefore, we were constantly forced to adapt our tactics in the face of difficulties, but – rather than seeing this as problematic – it is indicative of the 'messy' approach to politics that is integral to autonomous praxis (Chatterton and Pickerill 2010). The constant need to reflect on tactics is actually what prevents certain practices congealing into what radical food politics should 'look like' (Nunes 2005). Following Harris (2009), it is also important not to pessimistically read neoliberal logic into alternative food projects and, instead, explore "different readings" of these initiatives to see how they can 'open discursive spaces outside the perceived dominance of neoliberalism' (2009: 3).

In the context of the food giveaways, the performative sharing of food is what provides scope for this type of 'different reading.' Though giveaways share similarities with single-issue outreach, their realization – as performative resistance to existing uses of public space – enables vegan politics to be articulated in a more radical manner. Veganism's role in these protests is not simply the promotion of an alternative diet, but enacted as part of an explicit contestation of corporate power. In an *Indymedia* article about a food giveaway in 2010, for instance, I stressed these different elements of the protest:

> [D]espite superficially (and characteristically cynically) trying to re-brand themselves as an 'ethical', 'green' company – the same problems remain in relation to workers' rights, litter, unhealthy food, exploitative marketing aimed at children, animal welfare, and the general steamrollering of anyone who tries to get in their way!... The most important part of the day was the amount of people who approached us and wanted to have long and serious discussions about the reasons behind the protest. It was particularly refreshing to have groups of teenagers approach us and want to talk at length about the importance of considering how what you eat relates to so many other issues.
>
> ('eva g' 2010)

Aside from my (somewhat hyperbolic) criticism of McDonald's I felt it was important to emphasize the two aspects of the protests we found especially valuable: the way they drew together different forms of exploitation – resonating with Clark's aforementioned argument that "as a site of resource allocation, food tends to recapitulate power relations" (2004: 22) – and how they opened space for dialogue about these overlapping forms of exploitation.

It is here that the role of veganism, as an *enacted* practice, became particularly important. By cooking and serving vegan food as a direct counterpoint to McDonald's products, people were encouraged to sample these alternatives and engage directly with us. The fact that the food was vegan was especially significant, as it played an important role in inviting critical questions from the public about what we were doing and facilitating dialogue about existing human–animal relations.

Enacted in this way veganism does not have to function as an abolitionist imperative (even if many activists *personally* believe that animals should not be used for food), but instead can be used to unsettle categorizations of humans-as-consumers and animals-as-commodities that are ordinarily reinforced by fast-food restaurants. In this instance, cooking and serving vegan alternatives outside McDonald's directly disrupts the purchase of Big Macs, intervening in the particular set of relations that naturalize these categories on an everyday basis. The protests thus address Pedersen's argument (coming from a CAS perspective) that posthumanism should not just "rethink" but "*remake*" human–animal relations (2011: 74, italics in original). This goes back to the importance of challenging existing consumption practices on both a theoretical *and* a material level in order to disrupt the commodity-consumer relation and the categorization of humans (as privileged subjects) and animals (as exploitable objects) that is engendered by it. In opening dialogue about these categories – and intervening in practices that reinforce them – the protests, therefore, seem to bear out Pedersen's argument that veganism can be "one among other transformative moves" (2011: 75), which can facilitate this 'remaking' of relations.

From the perspective of nonrepresentational theory, Roe's work indicates further possible intersections between food giveaways and animal geographies. Roe contends that even simple acts, such as consuming a fast-food burger, necessitate specific affective environments (a complex assemblage of actors is involved in creating the burger's taste and promoting it as an appealing item of food, for instance). She suggests, conversely, that it is possible to transform these consumption practices by creating an "affective ethic that continues to materially connect the burger to the birth, killing, cutting-up and processing of an animal's body" (2010: 262). Roe herself focuses on how this has occurred within the assemblages of relations involved in meat production, due to animal sentience emerging as an actor that shapes the production process. Her example explores how this sentiency manifests itself in the quality of meat; because stressed animals produced meat with a higher pH level, improvements in welfare were seen as essential (Roe 2010: 275). Such transformations are deeply problematic from a CAS perspective, though, being due to commercial rather than political imperatives, with agency only manifested with the animal's death (in the 'poor meat' it produces).

The question that is informative in developing dialogue between these non-representational arguments and CAS is whether it is possible for *activists* to make the "affective connections" Roe outlines, and challenge existing human–animal relations without reinstalling representational rights frameworks. It is in this light that food giveaways are particularly informative. Roe herself suggests the performative preparation and cooking of food, in local food festivals for instance, can "present, discuss, (and) elaborate novel engagements with food through tacit, sensual, affective experiences" (2013: 3). She goes on to argue that, when such events actively involve would-be consumers in these perform-ances, they can offer "provocative, visceral encounters with food" that disrupt normative consumption practices through creating new relations with food pro-duction. Her argument is that, despite the temporary nature of these engage-ments: "one-off 'events' that mix the familiar with the strange and wacky, can be effective at punctuating the everyday and in so doing become the stuff of memories, informing without didacticism" (2013: 4).

These arguments resonate with existing theorizations of autonomous food politics, such as Sbicca's claim that, in "making-visible" a set of relations per-taining to food consumption, and involving the public in enacting alternatives, Food Not Bombs attempts to "mend metabolic rifts" that exist between urban dwellers and food production processes that are ordinarily rendered invisible (2013: 5). The performative serving of vegan food by Nottingham Vegan Cam-paigns can, likewise, be seen as a way of making the connections Roe describes, through altering people's bodily engagements with the spaces around McDon-ald's. The cooking and serving of veggie burgers in unexpected locations encourages people to actively smell, taste, and consume alternatives whilst simultaneously 'punctuating' the everyday spaces in which McDonald's burgers are normally consumed. In this way the protests do not simply create dialogue about the consumer–consumed relationship, or temporarily intervene in this rela-tionship by replacing veggie burgers for Big Macs, but alter the affective dynamic of the commercial locations in which these relations are enacted and reinforced on an everyday basis.

Conclusion

The decentering of the human called for by posthumanism and nonrepresenta-tional theory, which has proven so valuable for animal geographies, can be pro-ductively reconciled with CAS's call for concrete action by deriving conceptual insights from activism. Developing this conversation is important in overcoming both self-defeating essentialism and representational discourse, whilst actively contesting animal exploitation. It is also important in light of the need for both a conceptual and material challenge to existing human–animal intra-actions, to disrupt the categorization of animals-as-commodities and humans-as-consumers. Autonomous activism is a fertile source of insight, as it combines a desire to craft a concrete politics with a move away from representational practices. Its 'messiness' and constant self-reflexivity makes realizing this politics complex, but is also what makes it cosmopolitical and demonstrative of how the politics advocated by posthumanism and nonrepresentational theory could be – in Buller's terms – "operationalized."

In relation to food activism, both Food Not Bombs and Nottingham Vegan Campaigns demonstrate how existing social relations, which render certain actors 'exploitable,' can be contested within specific sites that naturalize and reproduce these relations on an everyday level. They also, however, demonstrate difficulties in realizing this politics and the need for constant reflection and adaption of tactics. In relation to food more specifically, despite the positioning of veganism as the established way of challenging the agricultural-industrial complex (a stance shared by CAS) which, seemingly, runs counter to autonomous values and posthumanist theory, its *enaction* in food giveaways demonstrates its more nuanced value. In these contexts it does three things to (in Pedersen's terms) remake human–animal relations: as a non-normative diet in the UK it opens space for dialogue about why activists are promoting it; in the serving of burgers, existing consumption practices are intervened in; and in the cooking and preparation of food affective relations are developed that reconnect animal bodies with the spaces in which they are consumed. Food giveaways, therefore, illustrate how complex and concrete contestations of animal exploitation can occur in practice, making this praxis informative for developing dialogue between posthumanism, nonrepresentational approaches, and CAS.

Notes

1 See Pedersen 2011 and Giraud 2013a for a more sustained defence of CAS.
2 There is obviously extensive debate surrounding the broader resonance of these claims; more specifically, McDonald's contested these arguments within the 'McLibel' trial. For nuanced analysis of the relations fostered by food corporations see Goodman and Watts (2005).
3 These issues emerged via internal email list discussions between activists who were attempting to organize an event at the Council House, but were told that community events were no longer allowed in the building due to 'wear and tear,' which prompted another activist to forward an email he had received about a commercial event that was taking place in the same location. When inquiring about the market square we were told it would cost £1,000 (with additional fees for marquee hire, electricity, and for any stalls we set up ourselves) and we could also only use specified contractors.

References

Adams, C 2006, 'An animal manifesto: gender, identity and vegan-feminism in the twenty-first century,' interview by T Tyler, *Parallax*, vol. 12 (1), pp. 120–128.
Anderson, B and Harrison, P (eds) 2010, *Taking place: non-representational theories and geography*, Ashgate, Farnham.
Barad, K 2007, *Meeting the universe halfway*, Duke University Press, Durham, NC, and London.

Bennett, J 2010, *Vibrant matter*, Duke University Press, Durham, NC, and London.

Best, S 2009, 'The rise of critical animal studies,' *Journal For Critical Animal Studies*, vol. VII (1), pp. 9–52.

Bird Rose, D 2012, 'Cosmopolitics: the kiss of life,' *New Formations*, vol. 76, pp. 101–113.

Braidotti, R 2013, *The Posthuman*, Polity, London.

Buller, H 2014, 'Animal geographies I,' *Progress in Human Geography*, vol. 38 (2), pp. 308–318.

Caffentzis, G and Federici, S. 2014, 'Commons against and beyond capitalism,' *Community Forum*, vol. 49 (S1), pp. 92–105.

Carolan, M 2011, *Embodied food politics*, Ashgate, Farnham.

Chatterton, P and Pickerill, J 2010, 'Everyday activism and transitions to postcapitalist worlds,' *Transactions of the Institute of British Geographers*, vol. 35 (4), pp. 475–490.

Clark, D 2004, 'The raw and the rotten: punk cuisine,' *Ethnology*, vol. 43 (1), pp. 19–31.

Collard, R-C 2014, 'Putting animals back together, taking commodities apart,' *Annals of the Association of American Geographers*, vol. 104 (1), pp. 151–165.

Crass, C 1995, 'Towards a non-violent society: a position paper on anarchism, social change and Food Not Bombs,' *The Anarchist Library*. Viewed November 12, 2014. http://theanarchistlibrary.org/library/chris-crass-towards-a-non-violent-society-a-position-paper-on-anarchism-social-change-and-food.

Dell'Aversano, C 2010, 'The love whose name cannot be spoken: queering the human-animal bond,' *Journal For Critical Animal Studies*, vol. VIII (1/2), pp. 73–125.

Despret, V 2013, 'Responding bodies and partial affinities in human–animal worlds,' *Theory, Culture and Society*, vol. 30 (7/8), pp. 51–76.

Dietz, T, Frish, AS, Kalof, L, Stern, PC, and Guagnano, GA 1995, 'Values and vegetarianism: an exploratory analysis,' *Rural Sociology*, vol. 60 (3), pp. 533–542.

DiVito Wilson, A 2013, 'Beyond alternative: exploring the potential for autonomous food spaces,' *Antipode*, vol. 45 (3), pp. 719–737.

'eva g' 2010, 'Vegan food giveaway at the AR spring gathering,' *Indymedia*. Viewed September 22, 2014. www.indymedia.org.uk/en/2010/03/447611.html.

Fox, N and Ward, KJ 2008, 'You are what you eat?: vegetarianism, health and identity,' *Social Science & Medicine*, vol. 66 (12), pp. 2585–2595.

Giraud, E 2008, 'McLibel to McSpotlight,' *E-Sharp*, vol. 12. Viewed September 12, 2013. www.gla.ac.uk/media/media_103039_en.pdf.

Giraud, E 2013a, 'Veganism as affirmative biopolitics,' *Phaenex*, vol. 18 (2), pp. 28–46.

Giraud, E 2013b, 'Beasts of burden: productive tensions between Haraway and radical animal rights,' *Culture, Theory and Critique*, vol. 54 (1), pp. 102–120.

Goodman, D and Watts, M 2005, *Globalising food*, Routledge, New York and London.

Greenhough, B and Roe, E 2011, 'Ethics, space, and somatic sensibilities: comparing relationships between scientific researchers and their human and animal experimental subjects,' *Environment and Planning D: Society and Space*, vol. 29 (1), pp. 47–66.

Guthman, J 2008, 'Bringing good food to others: investigating the subjects of alternative food practice,' *Cultural Geographies*, vol. 15 (4), pp. 431–447.

Haraway, D 1992, 'The promises of monsters: a regenerative politics for inappropriate/d others,' in L Grossberg (ed.), *Cultural studies*, pp. 295–337, Routledge, New York and London.

Haraway, D 2008, *When species meet*, University of Minnesota Press, Minneapolis and London.

Haraway, D 2011, 'Species matters, humane advocacy: in the promising grip of earthly oxymorons,' in M DeKoven and M Lundblad (eds), *Species matters*, pp. 17–26, Columbia University Press, New York.

Harris, EM 2009, 'Neoliberal subjectivities or a politics of the possible? Reading for difference in alternative food networks,' *Area*, vol. 41 (1), pp. 55–63. Viewed April 25, 2014.   www.communityeconomies.org/site/assets/media/EdHarris/Area-2009-Web-Version.pdf.

Heynen, N 2010, 'Cooking up non-violent civil-disobedient direct action for the hungry: "Food Not Bombs" and the resurgence of radical democracy in the US,' *Urban Studies*, vol. 47 (6), pp. 1225–1240.

Juris, J 2005, 'Social forums and their margins: networking logics and the cultural politics of autonomous space,' *Ephemera*, vol. 5 (2), pp. 253–272.

Kheel, M 2004, 'Vegetarianism and ecofeminism: toppling patriarchy with a fork,' in SF Sapontzis (ed.), *Food for thought*, pp. 327–334, Prometheus, Amherst, NY.

Kwan, S and Roth, LM 2011, 'The everyday resistance of vegetarianism,' in C Bobel and S Kwan (eds), *Embodied resistance*, pp. 186–196, Vanderbilt, Nashvile.

Latimer, J and Miele, M 2013, 'Naturecultures? Science, affect and the non-human,' *Theory, Culture & Society*, vol. 30 (7/8), pp. 5–31.

Latour, B 2005, *Reassembling the social*, Oxford University Press, Oxford and New York.

Lorimer, J 2013, 'Multinatural geographies for the anthropocene,' *Progress In Human Geography*, vol. 36 (5), pp. 593–692.

Lynn, WS 1998, 'Animals, ethics and geography,' in J Wolch and J Emel (eds), *Animal Geographies*, pp. 280–297, Verso, New York and London.

Mason, K 2013, 'Becoming citizen green: prefigurative politics, autonomous geographies, and hoping against hope,' *Environmental Politics*, vol. 23 (1), pp. 1–13.

McHenry, K 2012, *Hungry for peace*, See Sharp Press, Tucson, AZ.

Mitchell, D and Heynen, N 2009, 'The geography of survival and the right to the city,' *Urban Geography*, vol. 30 (6), pp. 611–632.

Nunes, R 2005, 'Nothing is what democracy looks like: openness, horizontality and the movement of movements,' in D Harvie (ed.), *Shut Them Down!*, pp. 299–319, Dissent! and Autonomedia, Leeds and Brooklyn.

Pedersen, H 2011, 'Release the moths: critical animal studies and the posthumanist impulse,' *Culture, Theory and Critique*, vol. 52 (1), pp. 65–81.

Philo, C 1995, 'Animals, geography and the city: notes on inclusions and exclusions,' *Environment and Planning D: Society and Space*, vol. 13, pp. 655–681.

Pickerill, J and Chatterton, P 2006, 'Notes towards autonomous geographies: creation, resistance and self-management as survival tactics,' *Progress in Human Geography*, vol. 30 (6), pp. 730–746.

Piper, N 2013, 'Audiencing Jamie Oliver: embarrassment, voyeurism and reflexive positioning,' *Geoforum*, vol. 45, pp. 346–355.

Portwood-Stacer, L 2012, 'Anti-consumption as tactical resistance: anarchists, subculture and activist strategy,' *Journal of Consumer Culture*, vol. 12 (1), pp. 87–105.

Robinson, A and Tormey, S 2007, 'Beyond representation? A rejoinder,' *Parliamentary Affairs*, vol. 60 (1), pp. 127–137.

Roe, E 2010, 'Ethics and the non-human: the matterings of animal sentience in the meat industry,' in B Anderson and P Harrison (eds), *Taking place: non-representational theories and geography*, pp. 261–280, Ashgate, Farnham.

Roe, E 2013, 'Enriching tacit food knowledges: towards an embodied food policy,' *Journal of Rural Studies*, (in press).

Sbicca, J 2013, 'The need to feed: urban metabolic struggles of actually existing radical projects,' *Critical Sociology*, doi: 10.1177/0896920513497375, pp. 1–18.

Shukin, N 2009, *Animal capital: rendering life in biopolitical times*, University of Minnesota Press, Minneapolis and London.

Smith, P 2009, 'Burgers not bombs,' *Peace News*, issue 2514, p. 1.

Smith, P 2011, 'President vegan,' interview with J Walker, *Left Lion*. Viewed January 5, 2014. www.leftlion.co.uk/articles.cfm/title/patrick-smith/id/3651.

Somdahl-Sands, K 2013, 'Taking-Place: book review,' *Antipode*. Viewed January 21, 2014. http://radicalantipode.files.wordpress.com/2013/08/book-review_somdahl-sands-on-anderson-and-harrison.pdf.

Stengers, I 2010, *Cosmopolitics I*, University of Minnesota Press, Minneapolis.

Stengers, I 2011, *Cosmopolitics II*, University of Minnesota Press, Minneapolis.

Thrift, N 2008, *Non-representational theory: space, politics, affect*, Routledge, New York and London.

Ufkes, FM 1998, 'Building a better pig: fat profits in lean meat,' in J Wolch and J Emel (eds), *Animal Geographies*, pp. 241–255, Verso, New York and London.

Urbanik, J 2013, *Placing animals*, Rowman & Littlefield, Lanham, MD.

Vegan Society 2014, 'World vegan month,' *Vegan Society*. Viewed January 30, 2014. www.worldveganmonth.net/events/running-a-stall-for-vegan-outreach/.

Veggies (n.d.), 'What's wrong with McDonald's?,' *McSpotlight*. December 10, 2013. www.mcspotlight.org/campaigns/current/wwwmd-uk.pdf.

Veggies 2014a, 'Vegan free food give-aways,' *Veggies*. Viewed January 30, 2014. www.veggies.org.uk/campaigns/vegan/free-food-givaways/.

Veggies 2014b, 'Vegan campaigning,' *Veggies*. Viewed January 20, 2014. www.veggies.org.uk/campaigns/vegan/campaigning-with-vegan-food/.

Veggies 2014c, 'History,' *Veggies*. Viewed January 20, 2014. www.veggies.org.uk/about/history-of-veggies/.

Vidal, J 1997, *McLibel: Burger Culture On Trial*, Pan Books, Chatham.

Warin, M 2011, 'Foucault's progeny: Jamie Oliver and the art of governing obesity,' *Social Theory & Health*, vol. 9 (1), pp. 24–40.

Weisberg, Z 2009, 'The broken promises of monsters: Haraway, animals and the humanist legacy,' *Journal for Critical Animal Studies*, vol. 7 (2), pp. 21–61.

Wilson, MW 2009, 'Cyborg geographies: toward hybrid epistemologies,' *Gender, place and culture*, vol. 16 (5), pp. 499–516.

Wilson, MW, Hickey, M, Craine, J, Fawcett, L, Oberhauser, A, Roe, E, and Warkentin, T 2011, 'Cyborg spaces and monstrous places: critical geographic engagements with Harawayian theory,' *Aether*, vol. 8a, pp. 42–67.

Wolfson, D 1999, *The McLibel Case and Animal Rights*, Active Distribution, London.

4 Pleasure, pain, and place

Ag-gag, crush videos, and animal bodies on display

Claire Rasmussen

If slaughterhouses had glass walls, we'd all be vegetarians.

The quote above, variously attributed to sources ranging from farmers to animal rights activists to conventional wisdom, expresses optimism about the ability of vision to transform our ethical sensibilities. The quotation suggests that only ignorance prevents a more moral position towards animals that takes account of the toll that our consumer practices take on the bodies of animals. Just seeing slaughter would presumably move us to think and act differently. This idea ironically motivates both the proponents and opponents of so-called ag-gag laws, measures currently being passed in states in the United States and under consideration in Australia. These laws prohibit the filming or photographing of animal production facilities, a common tactic of animal rights activists in exposing animal cruelty. Just as the activists rely on a sense of disgust to stir moral sentiments, proponents of these laws believe in managing representations of agricultural labor to avoid provoking such disgust. The political struggle here is one of managing affect: who has the power to control images of these animals and the presumed natural response to seeing them? Does distance from meat production prevent the kind of intimacy that might produce a different ethical relationship to our food?

But the politics of visibility are much more complex, as another legal struggle shows. So-called crush videos are pornographic films depicting animals being harmed, usually by women, for the purposes of sexual gratification. In the United States in 2010 a federal ban on the visual representation of cruelty to animals was struck down by the Supreme Court[1] as an overly broad restriction on the freedom of speech. Immediately, Congress passed a law narrowing the scope of the law to obscene depictions.[2] This other form of the legal regulation of representations of violence towards animal bodies depicts a very different relationship between affect and image in which the mobilization of desire, rather than disgust, is the object of regulation and political contestation. Here the intimacy between human and nonhuman animals is deemed inappropriate and disgusting.

In this chapter I argue that the connection made between vision and ethics – or seeing and caring for the Other – is problematic as a way of conceptualizing the political geography of human–animal relationships. The conflict over the meaning

– social, political, and ethical – of representations of animal pain demonstrates a deeper need to politicize rather than moralize our relationship with nonhuman life. The presumption that seeing leads to knowing and knowing leads to recognition reiterates many of the assumptions about subjectivity that underlie a liberal regime of rights. Remaining within the framework of the liberal subject of rights does not challenge the underlying assumptions that support a regime of violence towards animal bodies. What is needed is not an instance in which, as Derrida argues, "War is waged over the matter of pity" (2008: 29) but in which we confront animals as political co-subjects in all their complexity (Youatt 2012).

The legal regulation of vision

On the surface ag-gag laws and crush video prohibitions have very little in common. One regulates the ability to produce visual images of commercial farming operations, the other the production and sale of obscene materials that include harm inflicted on animals. The former is meant to protect commercial farming enterprises from animal rights activists; the latter is meant to protect animals. Both, however, seek to regulate the ability to see, and ostensibly the ability to respond to, the pain of/in animal bodies.

Ag-gag laws are a product of the American Legislative Exchange Council (ALEC), a pro-business conservative organization that drafts model legislation to be pitched in legislatures. While some laws date back to the early 1990s, the legislation gained momentum after 2008. The bills were driven in part by technological innovation, the invention of small recording devices, and the ubi-quity of camera phones that have made covert recording easier. The bills are also a response to several high-profile instances in which publicly disseminated videos of acts of animal cruelty led to significant legal action.[3] In 2008, for instance, a video produced by the Humane Society of the United States (HSUS) of a cattle operation in California led to the largest recall of beef products in U.S. history (Martin 2008). A 2008 tape of the Mowmar hog confinement facility in Iowa led to the first conviction for animal neglect against a Midwestern farmer, an event the facility's co-owner described as "the 9/11 event of animal care in our industry" (quoted in Genoways 2013).

The bills marked a shift in industry strategy from developing public relations responses to occasional leaks of images from their facilities to attempting to prevent the production of images themselves. So far nine states have adopted ag-gag bills.[4] In 2013, eleven bills were introduced. None has passed (HSUS 2014). The content of the bills varies from narrower measures that ban filming, sound recording, or photographing on private property without consent to broader measures making it a crime to gain access to property or apply for employment with the intent of producing images. Some versions of the legisla-tion include 'Good Samaritan' measures that punish individuals who see or record abuse but do not report it within a twelve-hour window, preventing the accumulation of evidence to show a pattern of abuse (Genoways 2013). Thus far, only one arrest has been made – of a woman who was recording activity outside of a slaughterhouse from a public road in Utah – and no convictions have been obtained under ag-gag laws (Potter 2013).

Legislative advocates of these measures have suggested that those who record in animal production facilities have less than savory motives, describing such recorders as terrorists who hold food producers "hostage ... to be persecuted in the court of public opinion" (Cohen 2014). In an even more over-the-top condemnation of animal rights activists, a legislator in Tennessee described the HSUS as a "fraudulent and reprehensibly disgusting organization of animal abuse profiteering corporatists, who are intent on using animals in the same way human-traffickers use 17-year-old women." He then went on to describe the filming of animal production facilities as "your preferred fund-raising methods ... a method that I refer to as rape and tape" (Hall 2013). The appropriation of the language of terror and rape – arguably the most violent terms available – is a forceful claim that to reveal the practices that go on within animal production facilities is, in itself, an act of violence, violating the private property of owners and their capacity to represent themselves.

Opponents of ag-gag legislation, while viewing the normative consequences of their actions quite differently, see the politics of visibility as equally forceful. While the constitutionality of ag-gag bills remains in question, the dispute over the regulation of filming animal production facilities demonstrates a faith in, and a fear of, the consequences of, as *New York Times* food columnist Mark Bittman describes it, making visible the "standard factory-farming practices" that make the production of cheap and abundant meat possible (2011). Activists seek to expose the everyday violence of meat production with the assumption that it presents a challenge to the acceptability of these practices.

While ag-gag laws suggest that disgust may be a natural outcome of seeing violence against animals, the existence of a need for laws prohibiting crush videos suggests quite the opposite: violence against animals can be aestheticized and eroticized in ways that are viewed as troubling. 'Crush videos' are a form of pornography in which animals are harmed for the purposes of sexual stimulation.[5] Most videos film a woman perpetrator either naked or partially naked as she commits the acts of violence, sometimes while also sexually stimulating another human or prior to sex with another human.[6] The initial crush video legislation was passed in 1999 to prohibit the production of images of animal cruelty.[7] In 2010, the Supreme Court heard a challenge to the law as a violation of the first amendment guarantee of free speech.[8] The particular case in question involved the prosecution of an individual who did not produce the original imagery but was producing and selling videos of dog fights and dog attacks on other animals. The court expressed concerns that Section 48, which defined the prohibited actions and noted exceptions to the law, was overly broad and could regulate a wide range of legal activities and/or stifle speech.[9] The court itself argued that the U.S. did not have a history of banning imagery featuring cruelty towards animals indicating that, unlike the historical regulation of obscene speech, it was not a form of speech believed to be harmful. Justice Alito offered the only dissent, arguing "the animals used in crush videos are living creatures that experience excruciating pain. Our society has long banned such cruelty."[10]

Congress and the president responded by passing and signing the Animal Crush Prohibition Act of 2010 which added a requirement that the images must also be obscene, limiting the prohibited images to those that were sexual in nature and thus not receiving first amendment protection.[11] The new language in the bill defines the criminalized images as follows.

(a) **Definition** – In this section the term 'animal crush video' means any photograph, motion-picture film, video or digital recording, or electronic image that–

    (1) depicts actual conduct in which 1 or more living non-human mammals, birds, reptiles, or amphibians is intentionally crushed, burned, drowned, suffocated, impaled, or otherwise subjected to serious bodily injury ... and

    (2) is obscene.[12]

The classification of a video as 'obscene' appeals to the general definition of obscenity as determined by the court as material primarily appealing to a prurient interest, being of a sexual nature, and lacking any redeeming educational, artistic, or social value. While presumably narrowing the range of speech being regulated to the constitutionally permissible arena of obscene speech, in the one case in which the federal government attempted to prosecute an offender, the court ruled against prosecutors because, the judge argued, the idea of animal cruelty being sexual was incomprehensible.[13] Judge Lake argued that in order to qualify as obscene the material must "portray sexual conduct in a patently offensive way ... but the acts depicted in animal crush videos may be 'patently offensive' under community standards, but under no set of community standards does violence toward animals constitute 'sexual conduct'" (quoted in Malisow 2013).[14]

Like the ag-gag legislation, the presumption in the regulation of crush videos is that the images produced prompt disgust, though unlike in the case of ag-gag videos that disgust is seen as a natural reaction to an unnatural action. The unthinkability of the eroticization of animal pain that manifests in the lower court judge's ruling indicates the sense that these practices are so abnormal and monstrous that the law cannot even recognize them. Here the source of disgust is the sexualization of pain. The law itself mirrors this differentiation between the normal and abnormal infliction of pain in carving out a set of exceptions to these images by designating:

(e) Exceptions –

    (1) **In general** – This section shall not apply with regard to any visual depiction of–

        (A) customary and normal veterinary or agricultural husbandry practices;

        (B) the slaughter of animals for food; or

        (C) hunting, trapping, or fishing.

Consequently, while in the case of ag-gag we see fear of denormalizing the everyday practices of animal production and slaughter, in the case of the regulation of crush videos the abnormal nature of these practices is highlighted by the desire to regulate not only the practices themselves but also the production of images of these practices. In both cases the fear is that the visibility of our intimate practices with animals will challenge our sense of humanity and propriety either by revealing the violence inherent in the operations of biocapital or in making visible the obscene and seemingly perverted practices that call into question the moral superiority of humanity.

The power of visibility

The desire to pass ag-gag and crush video legislation demonstrates a belief in the 'power of transparency' on both sides, believing that seeing and knowing the practices of violence against animals will be disruptive in itself. The anxiety provoked by crush videos and a desire to specifically ban the eroticization of animal suffering demonstrates that the relationship between power and vision is much more complicated. The act of seeing and the production of affect are not determinate and must be framed within a broader set of power relations that entangle humans and animals.

This argument is demonstrated strongly in Pachirat's (2011) *Every Twelve Seconds*, an ethnographic study of practices in slaughterhouses. He outlines the ways that distance, segregation, and concealment work to insulate the labor and production practices within the slaughterhouse from critique. The space of the slaughterhouse is privatized to not only avoid challenges to the practices inside but to allow society to 'forget' the practices that enable social reproduction. Thus the animal carcasses and the human labor become invisible as "a labor considered morally and physically repellent by the vast majority of society that is sequestered from view rather than eliminated or transformed" (Pachirat 2013: 11). Neither the animal bodies nor the bodies of the laborers, who tend to come from marginalized segments of the population, can be subject to moral concern or political challenge (see Joyce *et al.*, this volume). These dual concerns are captured in ag-gag laws in which the refusal to allow filming or pictures not only conceals the treatment of animals but the treatment of labor as well, as workers cannot document their own conditions. Of course, their vulnerability is magnified by the fact that, because vulnerable laborers – often undocumented workers – are those who have actual contact with animals, they are more likely to be prosecuted for animal cruelty than those who design policies or most directly profit from them.[15] Thus not only is the outside world prevented from seeing inside these spaces, incentives are created to remain opaque to the outside world. And even within the space of the slaughterhouse, labor is divided in such a way to prevent generating a full 'picture' of the process through segregating workers, dividing up different tasks, and other ways of concealing – even to those engaging in the labor – the full effect of their work (Pachirat 2013: 236).

Numerous studies have explored the variety of ways that the insertion of animals into the "animal-industrial complex" (Noske 1997) has worked to both apply the techniques of industrialized labor to the bodies of animals and produce animals as objects for human production (Stassart and Whatmore 2003; Ufkes 1998; Wolch and Emel 1998; Rifkin 1993; Emel and Wolch 1998). Concealment and distance play a role in this ability to transform animals into objects that are neither seen nor heard until they emerge as commodities. Yet even as Pachirat carefully unpacks the various ways that violence is concealed, he is also skeptical that unmasking is in itself a transformative process. He argues:

> [The politics of visibility] presumes a more or less standardized response, a generalized opinion, to industrialized slaughter made visible. Disgust, shock, pity, horror: the precise emotive label is less important than the assumption, the unarticulated expectation, of a reaction that would engender political action to end or transform the practices of industrialized killing.
>
> (2013: 247)

The conclusion here is twofold. On the one hand is a danger inherent in the reliance on affect that presumes a determinate emotional and political response to the visibility of violence, one that presumes that violence against animals or laborers will shock the conscience. The second is an unspoken narrative that the messy, material work of food production – its processes and its bodies – are themselves uncivilized and disgusting. This fails to interrogate the processes by which these forms of labor are stigmatized and the effect that has on the ability of workers or animals to be perceived as valued subjects and not as abject objects of derision.[16]

Thus, to understand the complex politics of visibility we have to take into consideration the ways in which affect is "not just the subject's private possession, but rather a socially operated inscription of who is sufficiently like us in the ways that guide deliberations over the institutional allotment of risk and protection" (Feola 2013: 138). In other words, to address both of the problems identified by Pachirat, we must elaborate the ways we socially produce our responses to violence while taking care to not reinforce the underlying forms of power that have produced those responses.

Adequately addressing the ways in which violence against some bodies is minimized requires moving beyond a naturalizing or biologizing presumption that relies upon what Wolfe (2010, 2012) calls ontologizing difference, assigning appropriate categories of being (subject/object, human/animal) that then determine the appropriate form of moral action toward that being. The previous discussion illustrates the ways that the clear distinction between human and animal, for example, is inadequate for these purposes since "the reproductive lives and labours of other species are also a matter of biopolitical calculation" (Shukin 2009: 12) and "modern biopolitics have forced themselves in the common subjection and management of both human and animal bodies" (Wolfe 2014: 162).[17]

Feola's examination of the concept of precarity in Butler (2009) is a useful starting point for thinking about the differential ways that affect is produced through the operation of power. The concept extends beyond just the ability to "make killable" (Haraway 2008: 80) or the "noncriminal putting to death" (1991: 110). The biopolitical operations at work involve a more complex network in which bodies are made to produce, to reproduce, or to die but in differential ways. All bodies may be subject to biopolitics but not all in a uniform way. Feola argues that exploring the precarity of some subjects requires a contextual exploration of the questions:

> [H]ow are lives exposed to death through sub-institutional processes? To what degree can a social mobilization of perception undermine a universalist moral discourse? How do forms of meaning and discourse render certain agents illegible as the kind of bodies that demand protection?
>
> (Feola 2013: 132)

Wadiwel (2002) asks many of these same questions in exploring the concept of "unnecessary suffering" that often defines legally permissible violence. If some suffering is unnecessary then some must be necessary, allowing us to inflict violence upon a nonhuman animal in particular situations (e.g., for research or for food). The concept of necessary or unnecessary suggests a complex calculation that is contingent, contextual, and highly political. Rather than fixing the human–animal boundary it settles it by deeming some forms of suffering acceptable on who suffers (migrant farm labor, domesticated farm animals), where they suffer (in public versus private spaces), and for what purposes (production of food versus sexual pleasure).

This differentiation between good and bad suffering surfaces in Alito's dissent in *Stevens*, the crush video case.[18] He strongly insists that society condemns cruelty but carefully carves out the acts depicted in crush videos and videos of dog fights from hunting, farming, or other activities. The argument on the one hand elevates a societal prohibition on cruelty that is evidence of a superior societal morality. Alito then turns around and differentiates acts of wanton cruelty from civilized violence such as hunting or farming that may have similar effects but are needless pain. Only some forms of violence are condemned in our long history of protesting violence against animals while others remain authorized within the tradition. The political parsing of pain is evident in the concerns about whether the law can adequately make a rational distinction between licit and illicit violence. Both the lower courts and the Supreme Court noted many of the apparent illogical consequences of a ban on depictions of cruelty, handwringing about whether the law would criminalize, for example, filming the capture of a fish out of season, or hunting a species that might be banned in another locale. While the court focused on the apparent legal conundrums created by the law, they do not address the ultimate source of the confusion: the continued political struggle to make sense of the human–animal boundary.

A more complex politics of visibility requires, then, exchanging an assumption that revealing the violence inherent in any particular practice will provoke a sense of responsibility or will lead to political action to transform those practices. Social processes that produce the meaning of particular bodies and of particular actions always already shape what we can see and our responses to it. We may not see or care about the suffering of some or may see it as necessary. Revealing actions against some bodies already deemed abject may reinforce their status as lesser. Thus a genuine politics of visibility must probe not only what is made invisible but how and why.

The subject of animality

If the first concern provoked by the politics of visibility related to how certain bodies matter and others do not, the second concern was whether the response to those bodies challenged or reinforced the underlying power dynamics that had rendered that body invisible. This section considers this question through a particular lens, the ways in which many conventional models of making animals matter – whether a model of rights or morality – reinforces a figure of subjectivity that has, in fact, supported the human–animal hierarchy. This concern mirrors the second concern of Pachirat mentioned above, that the terms in which we resist specific forms of power do not merely reiterate the terms by which violence was enabled. This section thus makes an argument for a specifically political rendering of animals that seeks to foreground the idea of power and the possibility of political struggle and community that does not presuppose the human. I suggest that political geography – as opposed to more familiar models of moral or political theory – is particularly well suited to this project.

One of the most familiar terrains for thinking about the 'problem of the animal' derives from moral and ethical philosophy that seeks to specify the criteria or capacities that can be used to designate animals as worthy of moral consideration. In these accounts, capacities such as sentience (Tom Regan) or a capacity to suffer (Peter Singer) are considered to be the morally relevant characteristics that qualify animals for moral inclusion. This move re-ontologizes animals, arguing that we can know the morally relevant characteristics and effectively evaluate animals as possessing those characteristics. These accounts become a "model of justice in which a being does or does not have rights on the basis of its possession (or lack of) morally significant characteristics that can be empirically derived" (Wolfe 2010: 75). These accounts, therefore, rely on a sense of ontological security about the ability to know – not only the nature and capacity of animals but also certainty about the nature of moral responsibility. In theory, this places the questions of what capacities are relevant and how they are determined beyond political contestation. Significantly, this model continues to place 'the animal' as the passive recipient of moral consideration rather than as complex beings who can interpret and act on their own worlds (Youatt 2012: 351).

In practice, the moral certainty of these philosophies often reinforces the superiority of the reasonable human.[19] Moral consideration of animals does not preclude the re-enactment of power relationships, specifically as the cultivation of kindness towards animals is configured as a feature that makes us more human and can justify dominion over others. Many arguments for kindness towards animals are based on a sense that these capacities make us more human (Boggs 2012: 35), allowing us to "transcend the beast within" or triumph over brutish behavior through the triumph of reason, a move which justifies the management or domination of others (Anderson 1997: 473). At the same time, this moralization of our relationship with animals has also justified the exclusion and marginalization of groups of humans perceived of as "inhuman(e)" (Elder *et al.* 1998).

Concretely, the discourse of welfare emerging from moral certainty often ironically justifies forms of violence towards animals. Agricultural practitioners often utilize the language of stewardship to describe a relationship of responsibility to the animals they raise for slaughter, highlighting a presumed cooperative network. This language obfuscates the relationship of domination and the extraction of capital from their animals (Ellis 2013: 443). Cole's (2011) discussion of the idea of 'happy meat' – animals we can feel good about eating because they have been humanely raised and slaughtered – demonstrates a similar logic by which the idea of humane treatment can encourage consumption.[20]

Some political theorists have also suggested that the current framework of liberal democracy can accommodate animals by simply revising our assumptions about animals in order to shift the terms of recognition and in doing so recognize and protect the fundamental worth of animals. The discourse of liberalism shifts the terrain: "The liberal state does not ask whether animals have souls, are inhabited by demons, or have been sent as punishment from God. Instead, it asks, what responsibility does the state have for their welfare?" (Smith 2012: 16).

Political theoretical accounts tend to draw analogies between animals and other excluded groups noting how they have been conventionally denied full autonomy, often because of a view of their embodiment or capacities. Animals are often compared with women, children or disabled humans as excluded because of their "enslavement to the body," cognitive (dis)ability, or other perceived lack (Wolfe 2010: 136). The conventional story proceeds, then, that animals can simply become a part of an expanded moral universe of subjects that has gradually included previously excluded groups.

Martha Nussbaum's *Frontiers of Justice* makes the precise comparison between disability and animality in order to make the claim that political justice does not require an assumption that the beings in question are "free, equal and independent" (2007: 87). Instead she draws on a capabilities approach that defines justice as chance for a "flourishing life" through respecting the dignity relevant to the specific being (or species). Nussbaum then turns to literature which, she argues, can provide a means of creatively identifying with our animal others and thus extending them respect and the opportunity for justice. She argues that literature may work to stir "in us identifications, empathetic responses, and projections that may then be readily formalized in analytical propositions" (2007: 411). The stirring of our sentiments, notably those of compassion, may spur us to recognize different kinds of animal dignity and thus develop new rules of conduct consistent with this moral recognition.

In critiquing this argument, Diamond points out that Nussbaum draws a straight line between the experience of compassion for an animal, or the provocation of affect via the medium of literature, to a rational ability to develop moral rules when

> what is 'indeterminate' is not our compassion for the suffering of nonhuman animals but the idea that 'rights' and 'entitlement' bear anything other than a completely continent relationship (derived from the historically and ideologically specific character of our juridical and political institutions and the picture of the subject of rights they provide) to the question of justice for nonhuman animals.
>
> (Quoted in Wolfe 2010: 79)

That is, Nussbaum, operating from an additive model of subjectivity, assumes animals may unproblematically be brought into the world of moral consideration without any modification to our basic assumptions about the nature of humans. We have merely committed a category error in the exclusion of animals. The critique here echoes, for instance, feminist arguments that bringing women into the moral and political universe of modernity does nothing to modify the basic assumptions that underpin their exclusion (Collard 2013). Nussbaum deliberately blurs the line between human and animal by suggesting that animals can be moral subjects and that the basis of our consideration is not reason but sentiment. However, she then clearly returns to a humanist model of justice in assuming that the ultimate response to this invocation of sentiment must be the crafting of better rules for human conduct. As with the moral accounts, this political model seeks to redraw but restabilize the human–animal boundary by suggesting that these questions can be resolved through human capacities (whether for reason or compassion) and thereby depoliticized (Rasmussen 2012).

Ultimately the moral and liberal accounts are inadequate in politicizing the categories of human and animal. While moving away from a strictly rationalist account, these theoretical accounts are concerned more with the ability to make animals compatible with existing systems than with the ways animals themselves may be transformative.[21] Oliver (2009: 22) makes this argument when she claims:

> We must reconsider our notions of autonomy and freedom in relation to animals and ourselves. Obviously the very conception of 'ourselves' or 'we' comes under scrutiny when we consider animals, not just because we may decide to include animals in those designations but also because we acknowledge that animals always have been formative parts of our self-conception, an avowal that necessarily transforms it.

The attempts to regulate representations of animals discussed above are resolved in the law, requiring a reference back to the subject of rights. In the case of ag-gag laws the conflict is drawn as between the rights of the property owner against the free speech rights of the activist-individual. In the case of crush videos the free speech rights of the individual are set against the ideal of the humane society. In no instances does the animal appear as a potential subject in its own right in a way that might demand "the repoliticization of animals as bodies and voices, not merely ideologies or conceptual tools" (Johnston 2008: 634).

The strategy of repoliticization requires a destabilization of all of the assumptions that have explicitly excluded animals from the realm of the politics because of their presumed inability to engage with 'man' as the political animal. To ask the question of how animals might be considered as active members of our political communities is to challenge not only the human–animal boundary but the boundary of the political itself. As Whatmore argues:

> [A]gency is reduced to the impartial and universal enactment of instrument reason, or 'enlightened self-interest,' institutionalized as a contractual polity of equivalent self-present individuals divested of difference context, or circumstance ... spatially and temporally stable conceptions of individual and collective social life—the sovereignty of the self and state—etched in the cartographies of the citizen and the nation.
>
> (2002: 148)

In short, thinking the political animal is to rethink the nature of politics itself.

Animal geographies, by virtue of studying engagements between humans and animals in concrete terms, have been particularly productive in thinking through the context-based and contingent nature of the relationship between humans and animals. Studies examining the biopolitical management of life have emphasized the ways that biopower sometimes erases differences between human and animal and at other times utilizes the distinction between forms of life to establish hierarchies and justify relations of domination. Thus, for example, Pachirat's description of the utilization of space in order to manage human responses to the spectacle of violence against animals is not entirely different from the now famous techniques utilized by Temple Grandin (1995) to shape the experience of animals in the production process through the use of the technological management of space and sensation.

The purpose of drawing these parallels between humans and animals is not to suggest that humans are no different from animals but to suggest that our existing models of political community that rely on a capacity for language, a contract of reciprocity, or shared recognition are inadequate for understanding the networks of power that entangle humans and animals. The agricultural spaces regulated by ag-gag laws are not best understood as a conflict between human and animal interests but as a complex set of relationships in which human and animal bodies are shaped by the demands of biocapital that establishes relationships of power between some bodies and others. The debate over crush videos is not a struggle to punish humans who behave as less than human or as inhumane.

Instead, the scene offers an opportunity to think about how some forms of violence are authorized and others are not in a complex entanglement of claims about humanity, sexuality, gender, and capital.

This transformed model of subjectivity is described by Wolfe as follows:

> We are forced to understand power, freedom, and resistance as modalities of responding to an other who is also taken to be able to respond, but it is a responding that takes place on the basis of forces and capacities that are in no way transparent, or even necessarily accessible to, the subject who responds. In this sense 'resistance comes first' precisely because it resides not just at the level of the body, 'before' the subject who takes thought, but also in the recursive relations of the body with its other—with all its others.
>
> (2014: 162)

Rather than an attempt to develop the rules of engagement by which we can decide once and for all the appropriate distribution of bodies in our networks of power, this idea of subjectivity suggests the need to think through the particularity of networks of power aimed at the direction of life in all its forms. This call echoes Urbanik's description of the expansion of animal geography to include "attempts to bring in the animals themselves as subjects of their own lives—whether part of ours or not—instead of just as objects of human control" (2012: 38).

The desire to frame the important issues in the ag-gag or crush video debate as a matter of human free speech or property rights versus animal welfare demonstrates the tendency to resolve complex political questions via a referent to a predetermined set of rules, in this case, the legal categories that take for granted the meaning of humans/animals, the structure of rights, and boundaries of moral consideration. The contribution of critical animal geographies can be to force us to examine the particular conflicts created by these taken-for-granted understandings. Looking at the conflict over ag-gag or crush videos from the perspective of the rights of the (human) subjects obfuscates not only the subjectivity of nonhuman animals but how the very subjectivity of humans is itself a product of their relationship with the human–animal boundary.

Conclusion

Bringing together two seemingly unrelated contemporary legal controversies involving seeing violence against animal bodies was an attempt to think through the complexity of the politics of vision. Metaphors of seeing or hearing others often draw on unproblematic assumptions that responsibility is located in *knowing* the other, obfuscating complicated issues of representation and power. While both sides of the ag-gag debate presume a politics of visibility that assumes a natural affective response of disgust to seeing violence, the case of crush videos suggests that the experience of seeing violence can be aestheticized or even eroticized. Both legal controversies draw our attention to our contingent and sometimes even contradictory responses to violence towards animals.

In the case of ag-gag, the presumption of both opponents of regulations on the depiction of violence and those who seek to use such depictions in activism, is that the everyday practices of animal production and slaughter would provoke revulsion if seen regularly. To the contrary, legislation against animal crush videos suggests and seeks to criminalize the eroticization of violence against animals. In doing so, lawmakers seek to carve out actions from which we derive pleasure from animals which ought to provoke disgust (eroticized violence) from the acceptable (hunting, fishing, animal husbandry, etc.). In both instances there is the suggestion that seeing violence may have a natural tendency to provoke a negative response and that response may indicate a normal sense of humanity that can differentiate between necessary and unnecessary suffering. The need for the law to make such a distinction, and its problems in doing so, indicates how these responses are in fact a product of power relationships that implicate not only a social construction of the human–animal boundary but also ideas about capital, sexuality, and even race and gender.

Our contingent responses to seeing the other suggest a need to understand the particular context within which violence takes place that frames our response. This contingency opens up the possibility of politicization, redefining not just our political community but what constitutes politics and who (or what) can engage. This suggests, as explored in this volume, that we need to seek more-than-human ways of producing knowledges about our more-than-human communities.

Notes

1 The case before the Supreme Court, *United States* v. *Stevens*, 559 U.S. 460, 130 S.Ct. 1577 (2010), involved a defendant convicted for the sale of dog-fighting videos rather than pornographic films, a fact used by the court to assert the overly broad nature of the ban.
2 As will be discussed below, the constitutionality of the replacement law is still in question.
3 While ag-gag legislation is a relatively new innovation, attempts to legislate the production of imagery associated with animal rights activism is not. Laws have targeted activists filming in facilities utilizing animal tests or those filming private individuals engaging in hunting. For examples, including "food disparagement laws," "animal enterprise interference," and "hunter harassment" laws see Perdue and Lockwood (2014), especially Chapter 3.
4 States that have adopted ag-gag laws include: Kansas, Montana, North Dakota, Iowa, Utah, South Carolina, Missouri, Arkansas, and Idaho (see Genoways 2013).
5 For media discussions of the practice, see Williams 2014, Thompson 2002.
6 These videos are distinguished from bestiality pornography that generally depicts a sexual act with an animal. Crush videos usually involve acts of physical violence towards an animal as a sexualized act. This distinction is tenuous since sexual acts on animals also often involve the infliction of pain.
7 See the text of the 1999 legislation here: www.gpo.gov/fdsys/pkg/PLAW-106publ152/html/PLAW-106publ152.htm (viewed September 8, 2014).
8 *United States* v. *Stevens*, 559 U.S. 460, 130 S.Ct. 1577 (2010).
9 The legislation was broadly supported by animal rights organizations who believed that exceptions for educational materials would protect the kinds of imagery produced

by their organizations (such as those prohibited by ag-gag legislation) in support of their cause. For an overview of the legal advocacy in the case see Perdue and Lockwood 2014, Chapter 6.

10 Significantly, this legislation only pertains to the production of images of crush pornography. Animal cruelty statutes still apply to the practices depicted.

11 The narrowing of the target of the law to obscene speech, historically considered unprotected, allowed the court and the government to sidestep a legal question some legal advocacy groups wished for the court to address: is the prevention of animal cruelty a compelling governmental interest that might justify limitations on free speech? The court has, for example, allowed limitations on the distribution as well as the production of child pornography on the grounds that discouraging the abuse of children is a compelling governmental interest that justifies limitations on production, distribution, and ownership of child pornography. The court and legislation, in highlighting the prurient nature of crush videos, avoided addressing the argument that prohibiting profit from depictions of cruelty toward animals would discourage such cruelty.

12 Full text of the bill available here: www.gpo.gov/fdsys/pkg/BILLS-111hr5566enr/pdf/BILLS-111hr5566enr.pdf (viewed September 8, 2014).

13 The case received significant media attention in a series of stories in the Houston Press News (Malisow 2013). The defendants produced a series of videos in which, among other things, the female defendant tortured a cat and decapitated a dog. They claimed that the animals were either their property or had been given to them by clients who asked them to produce crush videos for them starring the defendant.

14 This lower court ruling was recently overturned and federal charges under the Crush Video Act have been reinstated. The individuals involved in the video were also charged with animal cruelty under state statutes (Glenn 2014).

15 For a discussion of the vulnerability of laborers in agriculture see Shavers 2009, Stull and Broadway 1995.

16 Collard (2012) makes a similar argument in her critique of Haraway's insistence on the idea of an intimate ethics in which the face-to-face contact is a means of cultivating response and respect. This argument will be taken up in the next section.

17 The 'asininity' of the category of the 'animal' is unpacked in Derrida 2010 as he discusses how the binary of human and animal is itself a political move, one that mistakenly implies a homogeneity within each of two categories, ignoring the heterogeneity that exists within each term, while simultaneously insisting on difference between the two terms. See also Haraway's (2008) discussion of the scientific imprecision of the idea of 'species.'

18 See the full ruling here: www.supremecourt.gov/opinions/09pdf/08-769.pdf (viewed September 8, 2014).

19 For a more extensive discussion of this argument see Rasmussen 2011.

20 See also Evans and Miele 2012.

21 For an example of this see Kymlicka and Donaldson 2013.

References

Bittman, M 2011, 'Who protects the animals?,' *New York Times*, April 26. Viewed March 10, 2014. http://opinionator.blogs.nytimes.com/2011/04/26/who-protects-the-animals/.

Boggs, CG 2012, *Animalia americana: Animal representations and biopolitical subjectivity*, Columbia University Press, New York.

Butler, J 2009, *Frames of war: When is life grievable?*, Verso, New York.

Cohen, A 2014, 'The law that makes it illegal to report on animal cruelty,' *The Atlantic*, March 19. Viewed April 1, 2014. www.theatlantic.com/business/archive/2014/03/the-law-that-makes-it-illegal-to-report-on-animal-cruelty/284485/.

Cole, M 2011, 'From "animal machines" to "happy meat"? Foucault's ideas of discipli-
    nary and pastoral power applied to "animal-centered" welfare discourse,' *Animals*,
    vol. 1 (1), pp. 83–101.

Collard, R-C 2012, 'Cougar–human entanglements and the biopolitical un/making of safe
    space,' *Environment and Planning-Part D*, vol. 30 (1), pp. 23–42.

Collard, R-C 2013, 'Apocalypse meow,' *Capitalism Nature Socialism*, vol. 24 (1),
    pp. 35–41.

Derrida, J 2008, *The animal that therefore I am*, Fordham University Press, New York.

Derrida, J 2010, *The beast and the sovereign vol. 1*, University of Chicago Press, Chicago.

Elder, G, Wolch, J, and Emel, J 1998, '*Le pratique sauvage*: Race, place and the human-
    animal divide,' in J Wolch and J Emel (eds), *Animal geographies: Place, politics and
    identity in the nature-culture borderlands*, pp. 72–90, Verso, London.

Ellis, C 2013, 'The symbiotic ideology: Stewardship, husbandry, and dominion in beef
    production,' *Rural Sociology*, vol. 78 (4), pp. 429–449.

Emel, J and Wolch, J 1998, 'Witnessing the animal moment,' *Animal geographies: Place,
    politics and identity in the nature–culture borderlands*, pp. 1–24, Verso, London and
    New York.

Evans, A and Miele, M 2012, 'Between food and flesh: how animals are made to matter
    (and not matter) within food consumption practices,' *Environment and Planning D:
    Society and Space*, vol. 30 (1), pp. 298–314.

Feola, M 2013, 'Norms, vision and violence: Judith Butler on the politics of legibility,'
    Contemporary Political Theory, vol. 13 (2), pp. 130–148.

Genoways, T 2013, 'Gagged by big ag,' *Mother Jones*, July/August. Viewed September
    1, 2013. www.motherjones.com/environment/2013/06/ag-gag-laws-mowmar-farms.

Glenn, M 2014, 'Court reinstates "animal crush video" charges,' *Houston Chronicle*, June
    13. Viewed June 15, 2014. www.chron.com/news/houston-texas/houston/article/
    Appeals-court-reinstates-federal-animal-crush-5551952.php.

Grandin, T 1995, *Thinking in pictures and other reports from my life with autism*,
    Vintage, New York.

Hall, H 2013, 'Tennessee rep's e-mail calls Human Society methods "tape and rape",'
    The Tennessean, April 26. Viewed August 1, 2014. http://blogs.tennessean.com/pol-
    itics/2013/tennessee-reps-email-calls-humane-society-methods-tape-and-rape/.

Haraway, D 1991, *Simians, cyborgs and women: the reinvention of nature*, Routledge,
    New York.

Haraway, D 2008, *When species meet*, University of Minnesota Press, Minneapolis.

Johnston, C 2008, 'Beyond the clearing: Toward a dwelt animal geography,' *Progress in
    Human Geography*, vol. 23 (5), pp. 633–649.

Kymlicka, W and Donaldson, S, 2011, *Zoopolis: A political theory of animal rights*,
    Oxford University Press, Oxford.

Malisow, C 2013, 'Open season: Do laws against crush videos violate free speech?,'
    Houston Press News, May 15. Viewed July 1, 2013. www.houstonpress.com/2013-05-
    16/news/ashley-nicole-richards/.

Martin, A 2008, 'Largest recall of ground beef ordered,' *New York Times*, February 18.
    Viewed June 10, 2013. www.nytimes.com/2008/02/18/business/18recall.html?_r=0.

Noske, B 1997, *Beyond boundaries: Humans and animals*, Black Rose, Montreal.

Nussbaum, M 2007, *Frontiers of justice. Disability, nationality, species membership*,
    Taylor & Francis, Abingdon.

Oliver, K 2009, *Animal lessons: How they teach us to be human*, Columbia University
    Press, New York.

Pachirat, T 2011, *Every twelve seconds: Industrialized slaughter and the politics of sight*, Yale University Press, New Haven.

Perdue, A and Lockwood, R 2014, *Animal cruelty and freedom of speech: When worlds collide*, Perdue University Press, West Lafayette, IN.

Potter, W 2013, 'First ag-gag arrest: Utah woman filmed a slaughterhouse from the public street,' Green is the New Red blog. Viewed December 1, 2013. www.green-isthenewred.com/blog/first-ag-gag-arrest-utah-amy-meyer/6948/.

Rasmussen, C (2011), *The autonomous animal: self-governance and the modern subject*, University of Minnesota Press, Minneapolis.

Rasmussen, C (2012), 'Imagining otherness: The political novel and animal rights,' in A Sarat (ed.), *Special Issue: The Legacy of Stuart Scheingold, Studies in Law, Politics and Society*, pp. 155–179, Emerald, Bingley, UK.

Rifkin, J 1993, *Beyond beef: The rise and fall of cattle culture*, Plume, New York.

Shavers, AW 2009, 'Welcome to the jungle: New immigrants in the meatpacking and poultry processing industry,' *Journal of Law Economics & Policy*, vol. 5, p. 31.

Shukin, N 2009, *Animal capital: Life in biopolitical times*, University of Minnesota Press, Minneapolis.

Smith, KK 2012, *Governing animals: Animal welfare and the liberal state*, Oxford University Press, New York.

Stassart, PM and Whatmore, SJ 2003, 'Metabolising risk: food scares and the un/re-making of Belgian beef,' *Environment & Planning A*, vol. 35, pp. 449–462.

Stull, D and Broadway, M 1995, 'Killing them softly: Work in meatpacking plants and what it does to workers,' in D Stull, M Broadway, and D Griffith (eds), *Any way you cut it: Meat processing and small-town america*, pp. 61–83, University of Kansas Press, Lawrence.

The Humane Society of the United States (HSUS) 2014, 'Anti-whistleblower bills hide factory-farming abuses from the public,' *The Humane Society of the United States*, March 25. Viewed September 15, 2014. www.humanesociety.org/issues/campaigns/factory_farming/fact-sheets/ag_gag.html#id=album-185&num=content-3312.

Thompson, T 2002, '"Crush videos" plumb depths of perversion,' *Guardian*. Viewed April 1, 2014. www.theguardian.com/uk/2002/may/19/ukcrime.tonythompson.

Ufkes, F 1998, 'Building a better pig: Fat profits in lean meat,' in J Wolch and J Emel (eds), *Animal geographies: Place, politics and identity in the nature-culture borderlands*, Verso, New York.

Urbanik, J 2012, *Placing animals: An introduction to the geography of human-animal relations*, Rowman & Littlefield, Lanham, MD.

Wadiwel, D 2002, 'Cows and sovereignty: biopower and animal life,' *Borderlands*. Viewed September 15, 2014. www.borderlands.net.au/vol1no2_2002/wadiwel_cows.html.

Whatmore, S 2002, *Hybrid geographies: Natures cultures spaces*, Sage, London and Thousand Oaks, CA.

Williams, ME 2014, 'The dark world of animal crush videos,' *Salon*. Viewed September 15, 2014. www.salon.com/2014/04/18/the_dark_world_of_animal_crush_films/.

Wolfe, C 2010, *What is posthumanism?*, University of Minnesota Press, Minneapolis.

Wolfe, C 2012, *Before the law: Humans and other animals in a biopolitical frame*, University of Chicago Press, Chicago.

Wolfe, C 2014, '"A new schema of politicization": Thinking humans, animals, and biopolitics with Foucault,' in J Faubion (ed.), *Foucault now: Current perspectives in Foucault studies*, Polity Press, Malden.

Wolch, J and Emel, J 1998, 'Witnessing the animal moment,' in J Wolch and J Emel (eds), *Animal geographies: Place, politics and identity in the nature-culture borderlands*, Verso, New York.

Youatt, R 2012, 'Power, pain, and the interspecies politics of foie gras,' *Political Research Quarterly*, vol. 65 (2), pp. 346–358.

Part II

Intersections

5 Wildspace

The cage, the supermax, and the zoo

Karen M. Morin

Introduction

Standing at the top of the most iconic place on my university campus – the quad – one overlooks a beautiful pastoral landscape that encompasses the lower part of campus to the far extent of the Allegheny Mountains in the distance. This otherwise picturesque view of Pennsylvania's rolling hills and steepled architecture is always spoiled for me, though, by two equally visible sites of enclosure poking out of the landscape: the upper portions of the 'monkey cages' facility that is the hallmark of the university's Animal Behavior program, and the bell tower of the federal penitentiary on the edge of town (USP-Lewisburg). The extensive monkey cages facility currently contains an incredible array of animals for an undergraduate liberal arts school, including baboons, lion-tailed macaques, and squirrel and capuchin monkeys, all of whom live out their lives in small concrete lab cages. Few in the community would notice or care that these animals share a similar lived experience with the 1,000+ male inmates at the penitentiary, a site that has become notorious in recent years for its insidious Special Management Unit (SMU) program that features twenty-three- or twenty-four-hour a day double-celled lockdown of federal prisoners from around the country, brought here for two years of 'readjustment' (Morin 2013).

Much has been written about the penal philosophies, conditions, economics, and politics of caging humans in prison solitary confinement or lockdown cells (Haney 2008; King *et al.* 2008; Mears and Reisig 2006). Such scholars understand these practices as civil and human rights abuses. In a similar way, many animal rights scholars and activists challenge the conditions, ethics, and damage caused to animals confined in small cages in zoos and in other facilities with similar forms of captivity (Jamieson 1985; Kemmerer 2010; Acampora 2010). While I acknowledge that there are risks involved in making such comparisons and great care must be taken to do so (e.g., Spiegel 1996), a number of parallels can also be drawn between how humans and nonhuman animals alike experience and/or act upon such spatial tactics of enclosure.

This chapter brings forward a discussion of 'the cage' from a number of related angles for both human and nonhuman subjects and spaces. Comparisons between humans confined in maximum security conditions in prisons and jails, and nonhuman animals confined in cages in zoos and other zoo-like structures, offer a number of opportunities to study the relative histories, practices, experiences, and politics of caging. I consider the critical intersections of caging nonhuman and human bodies in American zoos and prisons, and the oppressions and structural inequalities that span species boundaries. These include the historical-geographical dynamics of the cage and underlying disciplinary regimes of the zoo and prison; the cultural and sociological 'mandates' of caging; the associated psychological-behavioral experience of being caged; and the political and ethical challenges to long-term captivity in cages that the respective animal rights and prisoner rights movements have brought forward.

In many ways the crisis of hyper incarceration, and the human rights questions posed by increased use of solitary confinement and/or permanent lockdown in maximum security prisons, map onto the development of the zoo and debates about caging animals for human entertainment and resource. This chapter explores and compares the movements and rights discourses surrounding each, examining their effectiveness, relative successes, and roadblocks. What, if anything, can be learned by cross-pollinating these discursive and activist fields in attempts to advocate for both human rights and animal rights?

I take up these questions in the sections that follow. Before proceeding I would note that there are, of course, many forms of caging animals and humans that I do not discuss, but that have broad relevance to my topic. Animals caged in biomedical and other research labs (such as the one at my university mentioned above), despite protections especially for primate animals, remain at a critical crossroads; and animals captive within the cycles of industrial agriculture experience some of the most horrific conditions of the cage. A number of scholars likewise point to comparisons that can be made between zoo cages and other types of human enclosures (asylums, camps, etc.; see Malamud 1998: 115–116; Watts 2000). Such studies are helpful in thinking through other ways that animal and human enclosures can be studied together. However, I confine my attention to zoo and prison cages because study of their similar geographies, disciplinary regimes, and rapid transformations over the past forty+ years provides an opportunity to examine their respective ethico-political challenges that can, in turn, speak back to theories that inform both critical animal and human geographies.

The caged body

The caged bodies of animals in zoos and humans in maximum security prisons can be compared through a number of rhetorical and material parallels. They both can be situated as spectacular, wild, and dangerous bodies that 'require' enclosure; as victims of the physical and psychological abuses of enclosure; as oppressed bodies wholly without social rights – as homo sacer – bare lives (Agamben 1998; Rhodes 2009); and as bodily capital accumulation strategies for both zoos and prisons (Harvey 1998; Nibert 2002), among others. As Malamud (1998: 117) succinctly observes, when we make the institutional parallel of the zoo to the prison it is always the body that is at issue: "the body and its forces, their utility and their docility, their distribution and their submission."

Space limits a thorough treatment of these many resonant parallels, but suffice it to say that particularly important here are the intersections that have to do with the social construction of the caged body as wild and dangerous. The isolation of the cage for both animals and humans is intended for the ostensible protection and safety of others. Caging humans requires producing them as animalistic first (Wacquant 2001). The cage connotes a beast lacking in self-control, a human (usually male) who is seen to behave like a dangerous, brutal, coarse, cruel animal (Malamud 1998: 114). We might imagine a presumed innocence of caged animals compared to the presumed guilt of caged humans – the latter 'got what they deserve' and thus do not inspire the sympathy of neglected or abused caged animals; yet the dangerousness, wildness, and potential violence of both are presumed and, indeed, often showcased.

Malamud discusses the ways that early zoos appealed to a taste for horror by stressing, and exaggerating, the savagery of their animals (1998: 106). Braverman, along with scores of others, argues that the most crucial assumption underlying the entire institution of animal captivity is the classification of zoo animals as wild and therefore as representatives of their unconfined conspecifics. "Take this assumption away, and you take away the raison d'etre of the zoo" (2013: 6, 8–9). Jamieson, though, in his classic 'Against Zoos' (1985), describes the profound denaturing that occurs at zoos; while animals may look like their wild counterparts and share the same genetic code, they lack the behaviors, skills, and awareness of those in natural habitats. Acampora describes zoos as thus producing a "generic animality" (2006: 104). Nonetheless, zoos unquestionably portray their animals as wild and untamed, and intimately related to those in the wild and thus sharing in their plight. In attempts to maintain the wildness of zoo animals, many zoos have ceased giving them Western names. Today they are given a number (like the prisoner), or an African name, further locating them as undomesticated. 'Timmy' and 'Helen' have become Mshindi, Kweli, or Mia Moja, names the public will presumably associate with 'wild' and distant Africa, even if they were born in the U.S. (Braverman 2013: 9).

Prison inmates themselves turn to animal imagery to express the dehumanizing effects of isolation and exposure in the prison. Many express shame and anger at being caged in view of others that position them like animals in a zoo. As one inmate whom Rhodes (2009: 197) interviewed in Washington described it, "if you choose to put these people in a box with nothing, what you're gonna get out of that is stark raving animals. I've seen animals produced in this very hall. People who have just lost their total cool." An Academy Award-nominated documentary, *Doing Time: Life Inside the Big House*, covered the humiliations and rank conditions at USP-Lewisburg, including some poignant scenes of guards referring to the place as a zoo and inmates responding by 'woofing' like dogs or wolves for the cameras, and then declaring, "we're dying" (Raymond and Raymond 1991). The cultures of many prisons reinforce this relationship. At Douglas County Correctional Center in Omaha, Nebraska, where I am conducting research on spatial design and violence, the unit where the most at-risk or unstable inmates are isolated is informally referred to as 'the zoo' by staff and inmates alike.

Racial difference is basic to much of the 'criminal as animal' rhetoric, and animalistic representations of Black men in particular originate in various social arenas. As Wacquant has argued, the wide diffusion of bestial metaphors for criminals in the journalistic and political field – "where mentions of 'superpreda-tors,' 'wolf-packs,' and 'animals' are commonplace" – in turn supplies a powerful common-sense racist explanation for the massive over-incarceration of African-American men, "using color as a proxy for dangerousness" (2001: 104).

We also see a tension between the visibility–invisibility of both caged animals in zoos and humans in prisons. To be viewed, to be something to look at, is an integral part of the caging process (Benbow 2000; Acampora 2006: 103–115). Prisoners, especially those isolated in maximum security, are hidden and secreted away, outside of public view, though at the same time they are sub-jected to constant, uninterrupted surveillance within prison walls. When their caged experiences turn into a spectacle for outside audiences – whether for media outlets, museum histories, or live performances – the association of the wild and animalistic is oftentimes reinforced (Schrift 2004; Turner 2013). Zoo cages, on the other hand – public spaces at the heart of the urban center – are designed *as* spectacle. "At the zoo, for visitors to have their (spectacular) fun, the wild animals must be kept on display" (Acampora 2006: 102). While the zoo requires such visibility, perhaps many or most animals are in fact actually hidden from view much of the time, when the zoo is closed and also when rotating on and off display (DeGrazia 2002; Braverman 2013). Ultimately we see a kind of violence and disempowerment in the perpetual visibility of the caged prisoner and zoo animal. And that visibility – combined with the social invisibility or secrecy surrounding the experience of human and animal subjects – is integral to their animalization and objectification.

Many linkages can also be made between the behavioral and psychological responses and experiences, human and nonhuman, to caged enclosure and cap-tivity. These include depression, despair, lethargy, stress, fear, shame, and even-tually anger, acting out, and violence. These responses have the tendency to reinforce preexisting assumptions that the enclosed body is wild, bestial, and savage, thus requiring caging.

Many zoo critics argue that confining animals in cages produces anxiety, sadness, neurotic behavior, poor hygiene, and suffering (Jamieson 1985; Acam-pora 2006, 2010; Francione 2000; Rudy 2011; Braverman 2013), even for those under the care of the best-intentioned zookeepers (let alone those who neglect or mistreat animals). While some animals existing in such captivity are well fed, disease free, and comfortable (DeGrazia 2002: 92), their living environment is boring and physically stifling, their "jailhoused bodies" never fully able to engage their physical and mental capabilities (Acampora 2006: 99–103). Zoo animals typically lack companionship, adequate exercise, and stimulation. As Jamieson argued, zoos can never hope to provide experiences that animals deserve – gathering their own food, living in social groups, behaving in ways that are natural to them (1985: 109). While Rudy observes that most zoo animals today "reveal no outrageous behaviors such as self-mutilation or obsessive behaviors such as spinning or pacing," she argues that this is likely because animals who engage in those behaviors are simply taken off of display (2011: 126). Conversely it is the case that most zoo animals today were born in captivity and thus many zoo proponents argue that the cage is a normal, comfortable experience for them (Zimmermann *et al.* 2007).

Similar human responses to caging in maximum security lockdown and/or isolation in prisons have been well established in the literature. Mears and Reisig (2006: 34) assert that the supermax differs from earlier prisons with lockdown 'holes' in that they do not aim, even at an ideological level, to reform inmates; rather, they are intended to break prisoners through isolation. In the supermax, prisoners eat, sleep, live, exercise, and die in their cells alone or with a cellmate. Haney (2008: 956), Mears and Reisig (2006), and King *et al.* (2008) among many others describe the oppressive day-to-day inmate experience: it is one of overall sensory deprivation, isolation and loneliness, enforced idleness and inactivity, oppressive security and surveillance procedures, and despair. Those not broken by the system may become more dangerous and mean. Violence or the constant threat of it is one guaranteed byproduct of the supermax, and mental illness the other (Haney 2008; King *et al.* 2008). The supermax "manufactures madness": "prisoners subjected to prolonged isolation may experience depression, rage, claustrophobia, hallucinations, problems with impulse control, and an impaired ability to think, concentrate, or remember" (Magnani and Wray 2006: 100).

There are also a number of parallels for how both human and nonhuman actors attempt to resist caging practices and enclosure, acting on their condition with gestures and acts (Rhodes 2009: 199). These are agents who oftentimes have no other means of resistance other than using their own bodies as weapons. As Braverman (2011: 1700–1701) argues, "despite their subjugated legal position, animals are nevertheless active subjects embodying a form of agency in their ability to continue to challenge, disturb, and provoke humans." Among other actions, animals can kick back at or attack their human caregivers and perhaps even exercise their own 'natural laws' (also see Philo and Wilbert 2000: 14–23). The recent documentary *Blackfish* (2013) testifies to the violent behavior of SeaWorld's orca whale Tilikum, produced as a result of prolonged captivity in small tanks. News stories and television programs such as the popular *Animal Planet* frequently play up the horrors and 'irrationality' of zoo animals attacking their human caregivers as such (e.g., Gates 2013).

Likewise, much could be said about the few opportunities that isolated prison inmates have to resist their spatial enclosure and associated punitive practices. Hunger strikes, refusal of cooperation, using their bodies as weapons, and creating violence are all tactics that carry numerous risks and ensure retaliation from prison administration. Throwing food and feces, acts of self-mutilation, biting, and so on are unsurprising outcomes of desperate individuals ensnared in the perverse and violent supermax system of punishment (James 2005). Such tactics and agency, while challenging the mute passivity of the imprisoned also, though, further reinforce preexisting associations of prisoner animalism.

This brief overview of some of the resonances across the caged bodies and experiences of zoo animals and prison inmates offers a useful context for examining the historical transformation of the zoo and prison as spaces of enclosure. I turn to those now, and subsequently to the various rights movements that have influenced the development of these spaces especially over the past forty years.

The zoo cage

Zoos are among the most popular cultural institutions worldwide; approximately 100,000 of them attract over 600 million visitors each year (Zimmermann *et al.* 2007: 4; Bulbeck 2010: 85). Today there are many kinds of zoos, and they vary considerably in quality and purpose, ranging from small, bleak 'concrete prisons' to naturalistic, conservation-oriented bioparks that attempt to replicate animals' natural habitats (DeGrazia 2002: 88; Rudy 2011: 122; Acampora 2010).

The zoo as a site for modern entertainment, education, or scientific study emerged in continental Europe in the eighteenth century, and in the UK and U.S. in the nineteenth. Hallman and Benbow (2006) explain the evolution of Western zoos in three distinct stages (also see Acampora 2005, 2006, 2010). The early menageries were dedicated to entertaining the public, with rows of bare cages enclosing single specimens and intended to reinforce "notions of human power and superiority over the natural world" in the age of colonialism and empire (Hallman and Benbow 2006: 257). Fast-forward to the postwar era, the "living museum" began to emerge, with enclosures built to resemble jungles and woodlands. Such spaces were meant to "banish the emotional response to human dominance over less powerful animals," emphasizing ecological relationships, habitat and species conservation, and public education. The late twentieth century "conservation centre" appeared as the third stage, a place that "exhibits active concern about the exploitative relations humans have with animals" and thus brings human visitors "inside the cage." Protecting biological diversity and sustainability are central to these later institutions (Hallman and Benbow 2006: 259–261). Samuels (2012: 33) adds a fourth stage of zoo development to this typology, a mode of display he characterizes as "eco-tainment," in which "giddy amusement-park tricks offer a measure of relief from the knowledge that nature is only another man-made illusion." Beardsworth and Bryman call this the "Disneyization" of zoos, involving a combination of "theming," consumption/merchandizing practices, and the emotional labor of zoo workers (2001: 91–98).

Such developmental stages can and often do coexist at any given site, as my recent visit to New York's Bronx Zoo confirmed. Widely acknowledged to be one of the world's 'best' zoos, the Bronx Zoo of today dedicates itself to "saving wildlife and wild nature," and many of its exhibits are framed with information about conservation and endangered species (such as the reproduction of a poacher's truck on Tiger Mountain). Nonetheless the Dinosaur Safari, Bug Carousel, and the Wild Asia Monorail, all aimed at having fun, obviously distract the zoo visitor, perhaps particularly children, from the more painful reminders of animal abuse and habitat demise.

It is clear that in the last forty+ years we have witnessed the emergence of an era of the benevolent or ostensibly 'progressive' zoo – an obvious reform of zoo conditions and a change of mission. Today, many zoo advocates argue that at least for larger primates and mammals, small concrete barren cages have generally given way to larger, more naturalistic habitat enclosures, some components of which are made of natural materials; animals are rarely displayed alone; and zoos attempt to educate the public about endangered species and habitat destruction noted in informational signage. Zoos vary tremendously on these features, however, and regulatory oversight as well as collection of empirical evidence to support such assertions has been unsystematic to date (Zimmermann *et al.* 2007; Braverman 2011, 2013; see below).

Debates about zoo cages today center on whether these progressive sites are in fact "a new and acceptable form of wild animal keeping, or whether they are simply a dressed-up version of the colonizing, concrete prison model" (Rudy 2011: 123–124; Acampora 2006: 103–115, 2010: 1–8). As Benbow (2000: 13–15) argues, technology and culture have conjoined in the development of new mechanisms for caging captive animals – including various forms of wire, glass windows, electronic fences, walls, ditches, and moats. She explains that the most significant change that has occurred in the geography of the cage is the (larger) size of enclosures. Nonetheless, in zoo architecture, a balance or compromise is always made between the conflicting aesthetic demands of visitors and the needs of the animals. Thus, rocks, vegetation, and other spatial features, for example, whether synthetic replicas or the real thing, are provided primarily for the visitor (not the captive animal), intended to evoke a natural environment, habitat, or themed region of the Earth (Benbow 2000: 18–20; Francione 2000: 24). Thus many would argue that the impression of 'cagelessness' in this 'natural' zoo habitat is merely pretense, with animals simply subjected to more sophisticated regulation (Malamud 1998: 107; Acampora 2006: 103–108). Ultimately, what all zoos have in common is the display of animals to the human public; and what all zoo animals have in common is the experience of being observed, as object, within a hierarchical relationship with the observer.

Moreover many critics argue that zoos do nothing to address the primary causes of global biodiversity loss, unless one considers the intangible benefit of education, which itself is highly debatable (e.g., Braverman 2013: 18). Many zoo critics such as Francione (2000: 25) argue that watching a lion in a zoo is no more beneficial than watching a film of a lion. Others, however, argue that witnessing the embodied "wonder, beauty and mystery" of zoo animals has great potential for changing public attitudes (Benbow 2000: 15, 2004: 379; Zimmerman *et al.* 2007: 4–7). As Jamieson (1985: 111–112) noted though, "undoubtedly some kind of education happens in zoos, but the question is, what kind? ... [C]ouldn't most of the important educational objectives better be achieved by exhibiting empty cages with explanations of why they are empty?"

Many also argue that zoos have not been successful at maintaining genetic diversity of endangered species and/or reintroducing species back into the wild, due to the substantial financial, health, and adjustment risks involved. Acampora (2006: 106) questions the utopian vision of "Releasement Day," arguing that this sort of futuristic planning presumes unrealistic, wide-scale, socio-ecological stability over a very long time; meanwhile Francione (2000: 24–25) describes the inefficiency, waste, and abuse inherent in breeding programs. Most zookeepers readily admit the obvious, that the best way to 'save' wild animals is to protect their habitats and sanctuaries (Rudy 2011: 119; Kemmerer 2010: 42). But because maintaining actual natural habitats and protected areas has not proven feasible, especially for large land vertebrates, the zoo becomes an (albeit controversial) 'necessity' to breeding endangered species and protecting biodiversity. Zoo advocates argue that modern zoos emphasize conservation, education, and care and stewardship towards animals as their central mission; that in caring for animals in zoos and breeding offspring, they are caring for animals in the wild: i.e., 'saving' wildlife (Braverman 2013: 15–17; Zimmermann *et al.* 2007).

A number of scholars and advocates argue for new models of wildlife enclosure altogether (the sanctuary, reserve, Earth Trust, zoological garden), and signal, if you will, a fifth stage in zoo development. Such models signal the abolition, rather than the reform, of what is ordinarily considered a zoo.[1] Rudy (2011: 113–114), for example, supports the privately run sanctuary, founded not on putting animals on display but on developing human relationships with them. Similarly Kemmerer (2010: 37–42) advocates for the "nooz," safe havens for individuals misused by zoos, circuses, or science, institutions framed within the logic of reparation for previous exploitation. "Nooz will not purposefully seek out prisoners from the wild, or breed prisoners to entertain human beings," she writes (2010: 42).

One such place is the largest exotic wildlife rescue facility in Pennsylvania, the family owned and operated T&D's Cats of the World near my home. Most of the 300 animals at T&D's – including lions, tigers, cougars, leopards, wolves, and bears – were formerly abused, neglected, or illegally owned pets, or have been discarded from zoos and other operations. Most of the animals live socially on a half to two acres of woods that also feature enclosed shelters. Although T&D's allows weekend visitors during the summer, it is a not a place designed for humans but rather for animals, the latter of whom may or may not 'display' themselves to visitors on any given day. T&D's is a good example of a facility that "contests exhibition" (Chrulew 2010: 205–206).[2]

All of this said and despite many disagreements it seems undeniable that a progressive social and spatial evolution has taken place over the last several decades in the politics, ethics, and care of animals in captivity – however much remains to be done. Real reforms are evident in the caged existence of zoo animals today, yet we also see a demonstrable further evolution, beyond reform of the zoo towards its abolition. What parallels can be drawn with the evolving practices of caging humans in prisons? Where do the narratives of these institutions converge, and where do they diverge?

The supermax prison cage

What we might think of as the social history of caging humans in long-term isolation can be traced in the U.S. to the infamous experiment at Eastern State Penitentiary in Philadelphia. Built in 1829, Eastern State is today considered America's most historic prison, primarily for the role it played in developing American penal philosophy. Philadelphia Quakers are attributed with creating the idea for this first penitentiary, a prison designed to inspire true regret, or penitence, in criminals' hearts through complete isolation, silence, and individualized labor in cells. Eastern State was a source of debate from the beginning – Charles Dickens was one of its earliest detractors, in 1842 – yet its ideals were not abandoned until 1913 when they collided with the reality of overcrowding. The prison did not close until 1971 though, and the site became a popular tourist attraction beginning in the 1980s (Bruggeman 2012).

From the late nineteenth century, social reformers had sought to ameliorate the deplorable conditions in American prisons – they were overcrowded, poorly ventilated, dark, unhygienic spaces where prisoners were kept in solitary cages regardless of their crime. Most were incarcerated for nonviolent crimes such as horse theft and counterfeiting; and later, under Prohibition (1919–1933), for producing, transporting, or selling alcohol, which dramatically contributed to the massive prison-building spree in the early twentieth century. When the Federal Bureau of Prisons (BOP) was created in 1930, its mission was to reform this system, to rehabilitate prisoners through education, vocational training, and recreation, an approach in line with contemporary 'scientific' penology.

This ideology of reform and rehabilitation suffered a short life span, however; by the 1970s, these principles completely lost traction within the prison bureaucracy and in the courts, and stood in stark contrast to the norms and practices of everyday life inside penitentiary walls (Richards 2008). Guard brutality, overcrowding, unsafe working conditions, infrastructural deterioration, and inmate civil rights challenges led to a breakdown in the ability of the BOP to control its facilities, and uprisings frequently occurred at many federal and state facilities. Prisons became increasingly violent places and dozens of guards around the country were killed and numerous lawsuits followed (King *et al.* 2008: 146; Richards 2008: 9–10; Morin 2013: 384). By 1975 the Bureau had abandoned its concept of rehabilitation, and by 1984 the U.S. Congress passed the Sentencing Reform Act, which abolished parole for federal prisoners, guaranteeing that they must serve at least 85 percent of their prison sentences. The failed War on Drugs conceived in the early 1970s, more than any other single cause, contributed to the hyper-incarceration trends we see today (2.4 million people behind bars; Alexander 2012).

The Bureau's response to the further problems that these legal 'remedies' predictably caused, such as intense overcrowding and increased violence within prisons, led to the caging practices we see today – use of solitary confinement and permanent lockdown as a primary method of prison control. Pure punishment and retribution became the norm. Today maximum security isolation or lockdown has become an ordinary and entrenched fixture in American prisons at all scales, replacing older forms of temporary prisoner segregation.

In the U.S. there are more than thirty high-security (federal and state) super-maximum prisons, confining approximately 200,000 prisoners, and the number is growing (Richards 2008: 17–18). 'Law and order' arguments tend to frame the need for the solitary cage in public discourse, as advocates often claim that such confinement of the 'worst of the worst' is necessary to stem violence and keep our streets safe. In reality, estimates are that only about 5–9 percent of those in isolation cages are locked down for a crime committed on the outside (Vanyur 1995); nearly all of them have been labeled as gang leaders and/or have been accused of committing crimes (such as assault) while incarcerated.

As one USP-Lewisburg inmate wrote about the situation there (Morin 2013: 393), "the prison provides only 55½ square feet for two grown men to live and be locked down in for 18 to 24 months straight." While the Lewisburg SMU's double-celling lockdown practices bring about their own set of stresses and violent outcomes of which few seem aware, the notorious federal ADX prison in Florence, Colorado, serves as a more familiar example of just how far the caging of humans can and has devolved. At ADX, prisoners are alone in their cells at all times, they recreate alone, and at no time come into contact with another human being, sometimes for years at a time. Cells are self-contained, a spatial design that maximizes security by ensuring that inmates rarely leave their cells. Cells are made of concrete walls, floors, and ceilings, and all cell furniture is made of reinforced concrete. Each cell has only a concrete slab bed, with a built-in storage shelf, concrete desk and seat. Each cell has its own toilet, sink, and shower. Each cell also has its own vestibule and two doors – an inner open grill that allows direct observation of the inmate upon entering the vestibule and an outer solid door that "prevents the inmate from throwing things or firing projectiles at staff" (Vanyur 1995: 92). Most services that prisoners receive are delivered to them electronically or through the small hole in the cell door.

Only recently are we beginning to see, from the mainstream media and from the corrections industry, challenges to the abuses inherent in solitary confinement, as well as more pragmatic arguments about the efficacy and economics of the practice. Raemisch (2014), for instance, Executive Director of Colorado state corrections, checked himself into a solitary confinement cell in order to better understand the abusive nature of the practice and the psychological damage it caused. He lasted only twenty hours, becoming "twitchy and paranoid," spending his time counting the small holes carved in the walls. Raemisch concluded that confining men to small solitary cages does not solve problems, "only delay[s] or more likely exacerbate[s]" them. Goode (2013), likewise, reported that the Mississippi Commissioner of Corrections "used to believe that difficult inmates should be locked down as tightly as possible, for as long as possible." But after a rash of violence at the state's super-maximum security prison in 2007, rather than tightening restrictions he loosened them, allowing inmates out of their cells each day, offering basketball, a group dining area, and new rehabilitation programs. In response, "inmates became better behaved. Violence went down" (Goode 2013). Ultimately an entire unit was closed, saving the state more than $5 million.

The devolution to permanent isolation or lockdown in U.S. prisons as normal, everyday practice can be attributed to a number of social and legal trends noted above. But 'cracks' in this system are beginning to show that might help ameliorate or reverse the trends. At this juncture we might question the relative impact that the prisoner rights movement has effected within this scenario, and compare it to that of the animal rights movement that has similarly challenged caging practices at the zoo.

Prisoner rights and animal rights: resonances and dissonances

One of the important things to notice when comparing animal rights and prisoner rights in America is the crucial changes that have overlapped in curious ways over the past forty years. Prior to that (and grossly generalizing for brevity's sake), the humanitarian basis of Progressive Era reforms led to improvements in both prisoner and animal welfare (Finsen and Finsen 1994: 27). New organizations pushed for improved treatment of animals; and new agencies governing the caging of humans (including the BOP itself) were created, based at least in theory on reform-minded principles. At the same time, while conditions arguably improved for animals caged in zoos, starting in the 1970s conditions dramatically deteriorated for humans caged in prisons. While the zoo and the prison are obviously different kinds of institutions, run under vastly different regimes of power, both have been historically subjected to pressures from outside activist organizations that have to a greater or lesser degree effected their reform. Yet advocates for change in both the prison and the zoo have drawn on similar ethical and biopolitical arguments about the cage as a disciplinary geographical space, and as such, offer a nexus of interests that can be put to productive use for both critical human and animal geographies.

While it is difficult to isolate a unified movement advocating for the rights of prisoners as an oppressed group, it is nonetheless possible to locate the origins of the idea of prisoner oppression and rights within the Civil Rights Movement generally, particularly considering that African-American and other minority men and women have historically comprised a disproportionately high percentage of the U.S. prison population (currently 70 percent; Alexander 2012). To Gottschalk, race was the "crucible" for the contemporary prisoner rights movement in the U.S.; the race question gave birth to the most powerful and significant prisoner rights movement in world, yet the U.S. was simultaneously a forerunner in the construction of the carceral state (2006: 165). These two factors are linked. Gottschalk (2006) offers one of the most comprehensive explanations for why the American carceral state did not encounter a more unified opposition, a question closely tied to explanations for why prison activism has a complicated, un-unified history with few successes.

To Gottschalk, the term 'prisoner rights movement' refers to a broad range of moral, political-economic, judicial, legislative, and cultural activities and institutions (2006: 165–196). Key to her thesis is that strong, well-organized activism inside prisons, particularly inmates' alignment with the Nation of Islam, the Black Panthers, and other New Left organizations, exposed the deep racism in American culture with which few outsiders would align themselves, particularly as their activism included strikes, uprisings, and calls for revolution. (Even the NAACP chose to focus more on affirmative action in schools and the workplace during this period, rather than on prison reform.) Some of these activists became household names, such as Malcolm X, George Jackson, and Angela Davis. In turn this activism created a strident 'law and order' backlash from conservative hard-liners, fed as well by the successes of victims' rights movements that perhaps unwittingly facilitated the more punitive turn in corrections and caging we see today.

Within this context the prisoner rights movement – if we can call it that – evolved as diffuse and frayed efforts, with agendas, tactics, and philosophies varying greatly from place to place and across many scales of activist organizing. Today, groups at various civic scales focus on a broad range of issues, from working to improve the conditions of confinement within prisons to abolishing prisons altogether. Agendas range from calling for sentencing reform and an end to mandatory sentencing; providing resources for legal representation; opposition to the death penalty; assistance to families of the incarcerated; self-help, vocational training, and education; religious rights of prisoners; and recidivism and re-entry. A number of national-scale organizations address prisoner issues, such as the American Civil Liberties Union, the Center for Restorative Justice, and the Sentencing Project; as do regional and state organizations such as Decarcerate PA and California's Mothers Reclaiming Our Children; as well as those more local and/or devoted to the rights of individual inmates such as Mumia Abu-Jamal's *Live from Death Row*. This short list does not begin to capture the vast number of organizations and coalitions active today – from grassroots and community-based groups to the more mainstream and institutionalized; from those based on prisoner organizing on the inside to those at work on the outside; and from scholarly and academic organizations to those based within the corrections industry itself.

Within this landscape are many organizations that have aspired to improve and reform conditions on the inside. To the extent that prisoner rights groups have attempted to effect tangible improvements of caging policies and practices over the past several decades – specifically those to do with the degrading practices of solitary confinement and permanent lockdown – they have done so mostly by bringing court action based on constitutional or civil rights of inmates, typically based on violations of the Eighth Amendment to the U.S. Constitution that guarantees freedom from cruel and unusual punishment. But these have had negligible success rate in the courts (ACLU 2013). Activists have to some extent begun to influence those within the industry though, such as the National Institute of Corrections, prison architects and planners, as well as various guard unions. Such entities recognize, among other things, that dehumanizing caging practices put staff at considerable risk, as well as create further instability among persons returning to life on the outside (Morton 2008; Raemisch 2014).

Turning to the zoo, the more recent transformations of its purpose and culture is a subset of the many changes in animal treatment, care, and use brought about most recently within the 'modern' animal rights movement, begun in the 1970s U.S. alongside other mid-century social movements. The idea that animals have intrinsic rights and/or selfhood resonated with these other movements (Jamieson 1985; Nibert 2002). The 1970s were also a decade within which animal suffering was witnessed on a new and grand scale (Finsen and Finsen 1994: 5–54). Animals became a much bigger part of medical and scientific research, and animal use in industrial agribusiness rose dramatically in the postwar period. With these as impetus, a virtual explosion of animal rights organizations arose in the 1980s U.S., with hundreds of organizations springing up at the local, regional, and national scales.

Perhaps unlike in other institutions within which animals are held captive and used as human resources (the farm, the research lab), transformations within zoos owed primarily to the ostensible 'voluntary' improvements made by the zoo industry in response to outside pressures from animal rights activists (Donahue and Trump 2006: 6; Braverman 2013). Zoos existed within the larger paradigm shift to animal rights and animal welfare, and in order to maintain (or establish) credibility, zoo keeping itself became professionally managed, with evolving cultural values, standards, and missions.

During the Progressive Era zoos were basically unregulated; zoo jobs were patronage jobs, and any improvements stemmed from pressure from local conservation or animal welfare groups. Beginning in the 1940s and 1950s, college-educated, conservation-minded biologists and zoologists began to take on roles as zoo managers (Donahue and Trump 2006: 8), and the professional organization, the American Association of Zoological Parks and Aquariums (AAZPA; later AZA, Association of Zoos and Aquariums), codified in 1972 as a separate organization. Subsequent changes to the organization co-emerged with the national animal welfare and rights groups which directly challenged zoo practices: the Animal Welfare Institute and the Society for Animal Protective Legislation among them.

As Donahue and Trump describe it (2006: 9), "despite their commitment to protecting wild animals, these professionals became formidable opponents" of animal rights activists and fought many political battles to continue operating zoos (also see Chrulew 2010: 195). Owing to outside pressures, the AAZPA was forced to clarify and (re)define its mission, eventually adopting those of conservation, education, and breeding vulnerable species. But lacking a singular purpose – and indeed attempting to sustain their conflicting and competing recreational, commercial, educational, and species preservation purposes – the industry also hired professional lobbyists to help align zoos with various conservation and animal protection groups. Faced moreover with congressional support for the federal regulation of zoos, the newly established professional network worked hard to ensure that "new laws provide zoos with a limited right to take protected animals from the wild, while at the same time acknowledging the independent regulatory capacity of accredited zoos" (Donahue and Trump 2006: 10–11, 37). In essence, zoos became and remain self-regulating institutions, subject to industry standards they themselves establish.

As Braverman (2011) explains, though, zoos are almost "extralegal creatures" regulated through myriad variances and exceptions. Owing to the powerful status of the AZA and "almost-monopoly" over relevant knowledge, zoo animals remain largely outside the provisions of official law (2011: 1703). As a result of "the physical and cultural nature of zoos on the one hand, and the long history of political battles between the zoo industry and animal rights groups on the other," a complex and eclectic mix of international, federal, state, and local agencies and codes regulate various aspects of zoo operations, but none address the zoo as a whole or the particular needs of most animals (Braverman 2011: 1694). Some only contain standards and care of warm-blooded animals; or of endangered species; or of particular kinds of animal breeding practices; or of only non-federally licensed operations; or have oversight only of building codes or of animals as property; and so on (Braverman 2011: 1689–1702). Indeed, most regulations governing zoo buildings are primarily aimed at protecting human accessibility and safety (Braverman 2011: 1700–1702; see note 2).

Thus the regulation regime of zoos, to the extent that one exists and influences practices, owes to a complex (though limited) bundle of self-regulating codes as well as various legal statutes at various scales. These in turn owe largely to much more widespread cultural pressures, including the more philosophical and ethical arguments that have led to reform of zoo caging practices as well as the abolitionist movement.

The critical nexus of human and animal geographies

Evidence suggests that the relative oppression and disenfranchisement of caged animals and prison inmates are closely linked, based on a range of policies, practices, and experiences. Since the 1970s in particular many scholars and activists have linked the structural and ideological similarities of racism, sexism, and speciesism (Nibert 2002; Francione 2008). As Watts categorically asserts, "[t]he zoo is a prison – a space of confinement and a site of enforced marginalization like the penitentiary or the concentration camp" (2000: 292). Many parallels and tensions can also be noted between the 'reformers' versus the 'abolitionists' with respect to both institutions: those who advocate for improving zoo conditions to those who wish to abolish them; and those who advocate for improved conditions of confinement within prisons to those who go beyond a discussion of civil or constitutional "rights" of prisoners to argue for the wholesale abolition of the prison industrial complex (Gilmore 2007).

At first glance it seems important to recognize the relative successes of animal rights activists in transforming the zoo compared to the relative ineffectiveness of prisoner rights activists to effect change in methods of incarceration. The zoo industry radically changed its mission, ideologies, and day-to-day practices owing fundamentally to animal rights activists whose strategies of evidence collection, public education, and savvy use of the media brought animal abuse issues to the forefront (Chrulew 2010). One of their key strategies was in publicizing to mainstream public audiences the horrors occurring in the secreted spaces of the animal cage. Lobbying efforts and notable legislative and judicial changes followed, and when they did not, organizations relied on protest and direct pressure to draw attention to animal abuses, including in zoos (Donahue and Trump 2006).

The potential impact of public intellectuals and journalists who have recently taken the same approach with respect to prison abuses cannot be overstated. Within the context of intensified use of control units, solitary confinement, and lockdown, and all their associated 'sort' response teams, heavy shackles, and myriad and severe daily restrictions that have become the norm, prisoners themselves have few options to mobilize, and those they do have carry enormous risks (Morin 2013). Thus some of the tactics and strategies successful with animal rights seem essential to shifting the grounds of the debate – that is, for advocates to bring issues directly to the public since legislative bodies and the courts have been unresponsive. Activists' attention to the questionable ethics of solitary confinement is beginning to make notable difference (cf. Gottschalk 2006), effecting change from within corrections that is not unlike the process by which the AZA fought but then eventually acceded to animal activists' pressures (e.g., Raemisch 2014).

Communication with and from inmates is essential in this process. One of the greatest misconceptions about prisons today is that the 'animalistic' behavior that occurs within their walls solely originates in the individual; that criminals are locked up because they are bad, deviant, or unfortunate people, driven to crime and trouble from some indelible social or psychological cause, and that their criminal nature will follow them wherever they go. Legislators, judges, educators, and journalists, as well as much of the public, are just beginning to understand the extent to which prison behavior and violence is a product of the carceral system, not an explanation for its need.

In many ways, though, the jurisdictional and regulatory regimes across types and scales of prisons are similar to what Braverman (2011) argues about the 'extralegal' nature of zoos and their self-regulation. While prisons reside within the American legal system and thus their practices are an instrument of it, super-maximum prison conditions also manifest features of a lawless, 'camp-like' space, exempt from outside scrutiny with inmate treatment typically beyond the scope of the law (Rhodes 2009; Agamben 1998). Prison administrators have been all too successful at making their own rules and keeping secret what happens on the inside. This, combined with the powerful influence of prison bureaucracies at every level, and community leaders with vested economic interests in keeping 'the machine' going, profoundly complicates the ability to make strident advances in prisoner treatment.

Acampora (2006: 109) meanwhile, citing the works of philosopher Heini Hediger, observes that there is a deep similarity between the professionalizing practice of zoo management and paternalistically progressive calls for prisons reform. In both cases, "one wants the inmates to feel as comfortable, as snug, and as much at home as possible." Such comfort perhaps belies nothing more than increasingly sophisticated means of regulating and disciplining captive bodies. While it is hard to deny that improvements in zoo conditions have occurred over the past few decades, the carceral comparison is not misplaced since both the zoo cage and the prison cage create the same result: an institution-alized organism largely incapacitated for life on the outside. Zoo critics argue that even the most sophisticated zoological garden, despite not carrying a puni-tive or penal intent (Bostock 1993: 63), is an institution of enforced occupancy intended to display animals. Like the prison, the structure of the zoo cage "ensures the production of docile bodies (or dead ones)" (Acampora 2006: 108–109). Again, it is visibility that is the key concern: the function of control-ling the animals' placement in the cage, controlling their diet, programming their breeding, and so on, "is to habituate them to tolerate indefinite exposure to the visive presence of humans.... [T]o display animals is already to discipline them" (Acampora 2006: 110, 112). Moreover, if released, they are subjected to a "carceral milieux": electronic tracking devices used to track animals are not unlike the parole boards (or tracking devices) of released prisoners (Acampora 2006: 110).

Prison abolitionists would agree. Davis (2003), James (2005), and Gilmore (2007), for instance, argue that reformers' rights arguments are misguided in that they reinforce the (illegitimate) power of the state to cage humans in the first place. James (2005) has collected many poignant prisoner stories framed around ethical arguments and the intrinsic human dignity of prisoners. Davis (2003) argues that the prison system is so judicially corrupt, racist, socially and econom-ically crippling to communities, and ineffective at reducing crime and keeping communities safe, that it is obsolete and should be abolished. And rather than focusing efforts on improving prisons, she argues that efforts should be redirected to crime prevention not punishment, education and support of vulnerable popula-tions, and community restitution for infractions, among others (Davis, 2003).

That said, some in the zoo abolition movement would see key differences between the "carceral milieux" of zoo animals versus prison inmates. For example, some would argue that because there are cognitive differences between human and nonhuman animals, humans are capable of responsibility and hence can be held culpable for their actions in ways that even the most sophisticated nonhuman animals should not be. So while we may observe equal vulnerability within zoo cages and prison cells, "susceptibility to incarceration" might remain invisible to our conscience (Acampora 2006: 108; Steiner 2008).

That complex issue will not be resolved here. But the ethical questions sur-rounding caged zoo animals and caged prisoners are clear. The incarceration tactics of long-term isolation and permanent lockdown that have become the norm in America today are wholly unjustifiable not least in that they have struc-turally targeted minority men in particular (Alexander 2012). For caged zoo animals, spectator interests and enforced display can and must be replaced by "loving sight" (Acampora 2006: 114). Ultimately, comparing the relative histories, practices, and biopolitics of the cage as a disciplinary geographical space offers us crucial insights to understand and then move stridently forward in challenging the abusive conditions that span species boundaries.

Notes

1 Other post-zoo models, beyond the scope of this chapter, call for an end to animal captivity altogether, in, for instance, the 'Zoöpolis' (Wolch 1998).
2 Nonetheless and quite obviously, the animals at T&D's remain enclosed and captive. Moreover, the facility is essentially self-regulating: while it is licensed by the U.S. Department of Agriculture and U.S. Game Commission, it is not accredited by the AZA – which, to its owners, would only mean a great deal of retrofitting for human visitors. While this particular sanctuary appears on the path towards zoo abolition, perhaps any number of animal abuses could occur at similar such facilities.

References

ACLU (American Civil Liberties Union) 2014, 'National prison project, prisoners' rights'. Viewed January 15, 2014. www.aclu.org/prisoners-rights.

Acampora, RR 2005, 'Zoos and eyes: Contesting captivity and seeking successor practices,' *Society & Animals*, vol. 13, pp. 69–86.

Acampora, RR 2006, *Corporeal compassion: Animal ethics and philosophy of body*, University of Pittsburgh Press, Pittsburgh.

Acampora, RR (ed.) 2010, *Metamorphoses of the zoo: Animal encounter after Noah*, Lexington Books, New York.

Agamben, G 1998, *Homo sacer: Sovereign power and bare life*, Stanford University Press, Stanford.

Alexander, M 2012, *The new Jim Crow: mass incarceration in the age of colorblindness*, New Press, New York.

Beardsworth, A and Bryman, A 2001, 'The wild animal in late modernity: The case of Disneyization of zoos,' *Tourist Studies*, vol. 1, pp. 83–104.

Benbow, SMP 2000, 'Zoos: Public places to view private lives,' *The Journal of Popular Culture*, vol. 33, pp. 13–23.

Benbow, SMP 2004, 'Death and dying at the zoo,' *The Journal of Popular Culture*, vol. 37 (3), pp. 379–398.

Blackfish, 2013, G Cowperthwaite, Director. Magnolia Pictures and CNN Films.

Bostock, S 1993, *Zoos and animal rights: the ethics of keeping animals*, Routledge, London.

Braverman, I 2011, 'States of exemption: the legal and animal geographies of American zoos,' *Environment and Planning A*, vol. 43, pp. 1693–1706.

Braverman, I 2013, *Zooland: The institution of captivity*, Stanford University Press, Stanford.

Bruggeman, SC 2012, 'Reforming the carceral past: Eastern State Penitentiary and the challenge of the twenty-first-century prison museum,' *Radical History Review*, vol. 113, pp. 171–186.

Bulbeck, C 2010, 'Respectful stewardship of a hybrid nature: The role of concrete encounters,' in RR Acampora (ed.), *Metamorphoses of the zoo:Animal encounter after Noah*, pp. 83–102, Lexington Books, New York.

Chrulew, M 2010, 'From zoo to zoopolis: Effectively enacting Eden,' in RR Acampora (ed.), *Metamorphoses of the zoo: Animal encounter after Noah*, pp. 193–220, Lexington Books, New York.

Davis, A 2003, *Are prisons obsolete?*, Seven Stories Press, New York.

DeGrazia, D 2002, *Animal rights: A very short introduction*, Oxford University Press, Oxford.

Donahue, J and Trump, E 2006, *The politics of zoos: Exotic animals and their protectors*, Northern Illinois University Press, DeKalb.

Finsen, L and Finsen, S 1994, *The animal rights movement in America: From compassion to respect*, Twayne, New York.

Francione, GL 2000, *Introduction to animal rights: Your child or your dog?*, Temple University Press, Philadelphia.

Francione, GL 2008, *Animals as persons: Essays on the abolition of animal exploitation*, Columbia University Press, New York.

Gates, S 2013, 'Tiger attack at Oklahoma zoo leaves worker injured after she sticks her arm inside cage,' *The Huffington Post*, October 6. Viewed September 15, 2014. www. huffingtonpost.com/2013/10/05/tiger-attack-oklahoma-zoo-arm-inside-cage_n_4050426. html.

Gilmore, RW 2007, *Golden gulag: Prisons, surplus, crisis and opposition in globalizing California*, University of California Press, Berkeley.

Goode, E 2013, 'Prisons rethink isolation, saving money, lives and sanity,' *New York Times*, March 10. Viewed September 15, 2014. www.nytimes.com/2012/03/11/us/ rethinking-solitary-confinement.html?pagewanted=all.

Gottschalk, M 2006, *The prison and the gallows: The politics of mass incarceration in America*, Cambridge University Press, Cambridge.

Hallman, B and Benbow, M 2006, 'Naturally cultural: The zoo as cultural landscape,' *Canadian Geographer*, vol. 50, pp. 256–264.

Haney, C 2008, 'A culture of harm: Taming the dynamics of cruelty in supermax prisons,' *Criminal Justice and Behavior*, vol. 35, pp. 956–984.

Harvey, D 1998, 'The body as accumulation strategy,' *Environment and Planning D: Society and Space*, vol. 16, pp. 412–421.

James, J (ed.) 2005, *The new abolitionists: (Neo) slave narratives and contemporary prison writings*, State University of New York Press, Albany.

Jamieson, D 1985, 'Against zoos,' in P Singer (ed.), *In defense of animals*, pp. 108–117, Harper and Row, New York.

Kemmerer, L 2010, 'Nooz, ending zoo exploitation,' in RR Acampora (ed.), *Metamorphoses of the zoo: Animal encounter after Noah*, pp. 37–56, Lexington Books, New York.

King, K, Steiner, B, and Breach, S 2008, 'Violence in the supermax: A self-fulfilling prophecy,' *The Prison Journal*, vol. 88, pp. 144–168.

Magnani, L and Wray, H 2006, *Beyond prisons: a new interfaith paradigm for our failed prison system*, Fortress Press, Minneapolis.

Malamud, R 1998, *Reading zoos: Representations of animals and captivity*, New York University Press, New York.

Mears, D and Reisig, M 2006, 'The theory and practice of supermax prisons,' *Punishment and Society*, vol. 8, pp. 33–57.

Morin, KM 2013, ' "Security here is not safe": violence, punishment, and space in the contemporary US penitentiary,' *Environment and Planning D: Society and Space*, vol. 31, pp. 381–399.

Morton, G 2008, 'Prison workers say they're at risk,' *The Daily Item*, July 8.

Nibert, D 2002, *Animal rights/human rights: Entanglements of oppression and liberation*, Rowman & Littlefield, Lanham, MD.

Philo, C and Wilbert, C (eds) 2000, *Animal spaces, beastly places: New geographies of human-animal relations*, Routledge, London

Raemisch, R 2014, 'My night in solitary,' *New York Times*, February 20. Viewed September 15, 2014. www.nytimes.com/2014/02/21/opinion/my-night-in-solitary.html.

Raymond, A and Raymond, S 1991, *Doing time: life inside the big house*, Video Verite.

Rhodes, L 2009, 'Supermax prisons and the trajectory of exception,' *Studies in Law, Politics, and Society*, vol. 47, pp. 193–218.

Richards, S 2008, 'USP Marion: the first federal supermax,' *The Prison Journal*, vol. 88, pp. 6–22.

Rudy, K 2011, *Loving animals: Toward a new animal advocacy*, University of Minnesota Press, Minneapolis.

Samuels, D 2012, 'Wild things: Animal nature, human racism, and the future of zoos,' *Harper's Magazine*, June, pp. 29–42.

Schrift, M 2004, 'The Angola Prison rodeo: Inmate cowboys and institutional tourism,' *Ethnology*, vol. 43, pp. 331–344.

Spiegel, M 1996, *The dreaded comparison: Human and animal slavery*, Mirror Books, New York.

Steiner, G 2008, *Animals and the moral community: Mental life, moral status, and kinship*, Columbia University Press, New York.

Turner, J 2013, 'The politics of carceral spectacle: Televising prison life,' in D Moran, N Gill, and D Conlon (eds), *Carceral spaces: Mobility and agency in imprisonment and migrant detention*, pp. 219–237, Ashgate, London.

Vanyur, J 1995, 'Design meets mission at new federal max facility,' *Corrections Today*, vol. 57, pp. 90–97.

Wacquant, L 2001, 'Deadly symbiosis: When ghetto and prison meet and mesh,' in D Garland (ed.), *Mass imprisonment: Social causes and consequences*, pp. 82–120, Sage, London.

Watts, MJ 2000, 'Afterword: enclosure,' in C Philo and C Wilbert (eds), *Animal spaces, beastly places: New geographies of human-animal relations*, pp. 293–304, Routledge, New York.

Wolch, J 1998, 'Zoöpolis,' in J Wolch and J Emel (eds), *Animal geographies: Place, politics and identity in the nature-culture borderlands*, pp. 119–139, Verso, New York.

Zimmermann, A, Hatchwell, M, Dickie, L, and West, C (eds) 2007, *Zoos in the 21st century: Catalysts for conservation?*, Cambridge University Press, Cambridge.

6 Commodification, violence, and the making of workers and ducks at Hudson Valley Foie Gras

John Joyce, Joseph Nevins, and Jill S. Schneiderman

The production of foie gras, French for 'fatty liver' (of a duck or goose), is highly controversial – especially among animal rights activists. Critics assert that its making involves inherently cruel practices. It is for this reason that California banned its production in 2012, joining more than a dozen countries – Denmark, Israel, Switzerland, and the United Kingdom among them – that have outlawed the culinary delicacy's production and, in some cases, its sale as well.

In the United States, much of the ire of the product's critics is focused on Hudson Valley Foie Gras (HVFG; e.g., Tepper 2013). HVFG is an industrial animal facility in New York's Sullivan County, an area formerly nicknamed the 'Jewish Alps' or 'Borscht Belt' for the now mostly defunct summer resorts frequented in the early to mid-1900s by middle- and working-class Jewish New Yorkers. The company is responsible for over half of the foie gras produced in the United States (Sella 2005). An HVFG spokesperson claims that the company controls about 70 percent of the U.S. market (Henley 2011, personal communication, October 28). While HVFG presents itself as a company open to visitors and public scrutiny, it exercises tight control over its workforce, and makes a concerted effort to limit access to, and critical examination of, its space of production – both forms of restriction aided by its rural setting.

Located roughly 100 miles north of Manhattan, HVFG grew out of the work of Izzy Yanay, an immigrant from Israel. Prior to coming to the United States, Yanay served in the Israeli military (IDF), received a bachelor's degree in agriculture, and was a field manager for a large Israeli foie gras producer. According to the HVFG website, Yanay is widely recognized as one of the world's foremost authorities in the breeding, growing, 'feeding' and 'processing' of ducks for foie gras production, having started the first U.S. foie gras 'farm' and processing plant in 1982. The operation was unique, so goes the official story, because it integrated all facets of the 'production process' in one location (Hudson Valley Foie Gras 2014).

In 1989, Michael Aeyal Ginor joined Yanay to establish HVFG, which touts a "unified, controlled and consistent operation based on nature, nurturing, and computer technology." Born to Israeli expatriates living in Seattle, Ginor has an MBA, worked for four years on Wall Street and then moved to Israel to serve in the IDF. In Israel, he also learned about foie gras processing. Thereafter, according to the HVFG website, he "aggressively" pursued his dream to establish "modern-age" and "total comprehensive production" of foie gras, which led to his partnership with Yanay (Hudson Valley Foie Gras 2014).

While HVFG avers that "the essence of farming is caring for animals," the company processes roughly 250,000 ducks annually by enlarging their livers through 'gavage.' This process entails force-feeding by inserting funnels and tubes down the birds' esophagi, thus allowing the 'feeder' to control the amount of food and frequency at which ducks 'eat.' Although HVFG insists that the form and function of duck necks are 'naturally designed' to accommodate the consumption of excessive quantities of 'food' without causing pain or distress, others contend that the large amount of food which is rapidly intubated during the force-feeding procedure leads to immediate esophageal distension, increased heat production and panting, and production of semi-liquid feces (European Union 1998). Though challenged by the American Medical Veterinary Association, the 1998 European Union Scientific Committee Report concluded that force-feeding was detrimental to the welfare of birds (De Rilke 2008: 91). Nonetheless, until late 2013, HVFG characterized its foie gras as "humane," only agreeing to remove the term from its website after the Animal Legal Defense Fund sued HVFG for characterizing foie gras in this way (Guillermo 2013).

Approximately 140 Mexican immigrant laborers employed at HVFG year round do the work of gavage, in addition to slaughtering and 'disassembling' – that is, dismembering – the ducks, and packing them for sale (Greenhouse 2001). They work long hours in an environment with a variety of health hazards, and often do not get a day of rest, having to work seven days a week. HVFG pays these workers the minimum wage allowed by the law, and provides no overtime pay or sick days (Herbert 2009). Such working conditions have also led to criticism (e.g., Greenhouse 2001; Herbert 2009), albeit considerably less than that brought about by the treatment of the ducks. What both the duck-rights-inspired and worker-rights advocates share, however, is a tendency to focus on the mistreatment of one kind of animal at HVFG, the nonhuman or the human; thus, both contingents of advocates have neglected to consider how the abuse of human and nonhuman beings at HVFG might be dynamically interrelated. In this regard, we share the perspective of Greta Gaard and other ecofeminists that "speciesism functions like and is inherently linked to" other processes of domination and subordination such as classism, racism, and sexism (Gaard 2002). These links unfold within and beyond the space of HVFG given that places are unbounded and are the geographical expressions of networks of dynamic social relations (Massey 1993, 2005).

Our particular case study highlights these socio-spatial networks, despite the tendency in the popular narratives about foie gras to ignore them. Our project stems from a late 2009 dinner attended by John that celebrated the efforts of former employees of HVFG to organize – efforts that ultimately led to their dismissal. How was it, he wondered, that folks so concerned about justice could celebrate the rights of workers but ignore the nature of the work they did as HVFG employees? Furthermore, how could a movement that endeavors to bring to light the wrongs embedded in a plate of food – in the form of injustice towards farmworkers – ignore the nonhuman beings whose flesh is also on the plate (and the violence that brought them there)? Laying bare these connections provides openings to address injustices against human and nonhuman animals, while assisting "in building alliances with others, especially those residing in different cultural contexts" (Deckha 2012).

Though much literature exists concerning nonhuman animal-for-food agriculture, there is very little academic work about ducks and foie gras, and none about a foie gras industry that depends on and reproduces relationships of violence and commodification experienced by both human and nonhuman bodies. This chapter explores those interrelationships, arguing that the mistreatment of ducks and workers in the foie gras industry is a form of objectification, one that reduces living beings in various ways to things or inanimate entities. It is a reduction brought about through a process of commodification that is inherently violent (broadly conceived). In illustrating this, the chapter demonstrates the complicated intersections of gender, race, national citizenship, mobility, species, and space. It also demonstrates the complicated geographies of commodification, and how HVFG relies both on mobility of the 'free' and coerced sorts as well as on immobility – of ducks and workers – and on particular geographies given the coproductive relationship between mobility and space (see Cresswell 2006). In this sense, HVFG is the geographical expression of intersecting trajectories that transcend simple territorial categories, while simultaneously depending on and helping to produce bounded spaces.

Producing people and ducks as commodities

While no doubt humans and ducks differ in significant ways, they also share much. In a world of 'isms' of domination and subordination, what they share can sometimes take the form of violence. Among humans, the violence flows from and produces unjust hierarchies along axes of difference such as those associated with class, nation, race, and sex (ones that sometimes involve animalization of 'others'). In the case of the systemic violence by humans against nonhuman animals, it is justified through a lens of speciesism and the concomitant assignation of a differential status. From a more egalitarian perspective, humans and ducks are both sentient beings. They feel pain, are capable of compassion, emote, and think. In the context of HVFG, both ducks and human workers are also commodities, ones whose (re)production involves violence.

Although we might consider a commodity as something that can be bought or sold, what it constitutes beyond this superficial observation is far from a simple matter (see Nevins and Peluso 2008). As Marx pointed out, a commodity is a "mysterious thing" (Marx 1992). In making this assertion, Marx was arguing against the reduction of a commodity to a lifeless product for purchase or sale, one whose value is determined by the price it fetches on the market, and for an appreciation of how commodities embody relations – social, geographical, and historical. They are relations that must be unveiled given how capitalism facilitates the obscuring of that which goes into the making of any commodity: by privileging exchange value, by reducing the value of a commodity to its monetary price, capitalism helps to hide the social and ecological relations involved in the commodity's production and consumption. In other words, Marx's project is to bring commodities to life by situating them in an ever-changing constellation of relations or context, to expose the inputs and work that makes them possible and the work they do (see also Appadurai 1986; Collard 2014; Nevins and Peluso 2008).

Given that the constellation of relations – in addition to idiosyncratic contingencies – surrounding any particular commodity is dynamic, commodities are always 'alive' in the sense that they move in and out of their commodity status; they have "social lives" (Appadurai 1986; Collard 2014). Of course, in talking about "lively commodities" such as people and nonhuman animals, they are never only commodities (see Collard and Dempsey 2013). This is because, for among other reasons, they often resist the conditions of commodification – as when workers, for instance, demand greater liberties and benefits – and thus engage in what we might consider practices of decommodification. This, in turn, invites responses by the forces of capital to re-commodify. For such reasons, we might think of animals and things existing along a spectrum, the endpoints of which are non-commodity and commodity. In other words, there are various levels of commodification.

Varying levels of violence underlie the commodification process. This is clearly so in the case of the production of ducks and foie gras and the related human labor – and not simply in terms of "primitive accumulation" (see Marx 1992) or "accumulation by dispossession" (see Harvey 2003; Glassman 2006). What Marx (1992) called the "real subsumption of labour" basically dictates, within the context of capitalism, the conditions of labor to such a degree that workers ("living labour") are subordinated to the needs of the production process, becoming a living part of that process. While Marx focuses on human labor, his analysis certainly illuminates what happens to ducks in the context of foie gras production. Through subsumption, they, too, become part of the production process, both subject to its dictates and helping to shape it, subject to and embodying its violent reductionism and participating in the making of violence (Lindner 2014, personal communication, January 31).

Here, the work of Johan Galtung is very helpful. Taking an anthropocentric approach in concerning himself with the well-being of humans, he argues that violence is "the difference between the potential and the actual" (Galtung 1969: 168). Moreover, it is made up of the "avoidable insults to basic human needs, and more generally to *life*" (Galtung 1990: 292, italics in original). In other words, what constitutes violence is that which operates against an individual reaching a full life – full in terms of both what is possible and, in the case of human beings, what most people in the world consider as desirable (Galtung 1969). Galtung is concerned here not only with events of physical violence, but also with the web of social relations that deprive all too many of the ability to have their basic needs met. In asserting his definition, Galtung posits three types of violence: personal or direct; structural or indirect; and cultural (what we call representational herein). When an identifiable actor commits violence, it is direct or personal. When there is no actor present – or when violence is the outgrowth of the seemingly acceptable, institutionalized practices of organizations deemed as legitimate, when it is part of the 'normal' social fabric – it is indirect or structural. Representational violence is that which serves to hide or justify direct and structural forms of violence, in Galtung's words, it "makes direct and structural violence look, even feel, right – or at least not wrong" (Galtung 1990: 291).

Speciesism is the principal representational violence that underlies the subordination and commodification of ducks. Speciesism flows from and produces a notion of human exceptionalism: the idea that humans have something of value that nonhumans do not, and that this 'something' legitimates disregard for nonhumans (Gruen 2001: 4–5). It thus entails "a failure, in attitude or practice, to accord any nonhuman being equal consideration and respect" (Dunayer 2004: 5). That attitude takes, among other forms, that of a 'big story,' one about a hierarchy of life at which human beings are at the top, thus conferring to humans a right to dominate nonhuman animals. And to the extent that it is a story that is so widely accepted, so hegemonic that it does not even need to be told, as it is simply common sense, the representational violence overlaps to a significant degree with structural violence in that it is invisible in some ways. As Galtung (1969: 173) explains in distinguishing between personal and structural forms of violence, the former "represents change and dynamism—not only ripples on waves, but waves on otherwise tranquil waters. Structural violence is silent, it does not show—it is essentially static, it *is* the tranquil waters." So, too, are hegemonic stories – such as that of evolution as a tale of progress in which humans are the ultimate form of biological development.

Of course, given the considerable degrees of direct violence involved in the commodification of ducks for purposes of foie gras production – after all, the ducks are ultimately killed – the speciesist story can only do so much work in obscuring that violence. In the case of structural violence – again, drawing on Galtung – it remains invisible only as long as the waters are tranquil – in other words, as long as widespread ideological conformity remains. As he points out, structural violence can become visible in a highly dynamic society (Galtung 1969: 173), implicitly one in which political forces effectively challenge dominant ideas of what constitutes violence and nonviolence. In the case of foie gras, those 'waters' are far less tranquil than they once were – an outcome that speaks to the power of organized activism. Given the 'luxury' and elite nature of foie gras (which is thus widely perceived as commodity of choice – for the privileged few – rather than of necessity), such an outcome also perhaps helps to explain the relative difficulty of delegitimizing other forms of commodification involving violence against nonhuman animals. In relation to, say, the slaughter of pigs for purposes of meat consumption, the violence is far less visible, in part because that which is consumed is deemed to be a basic need. As such – and one can say the same for all forms of violence that are seen as effectively less-than-violence – the violence is perceived as legitimate or necessary.

Violence of various forms – e.g., the taking away of the means of subsistence – is what produces a human population 'free' to work as wage laborers, and violence is what transforms land owned in communal fashion into something privately held and sellable to others. In this regard, we might say that the making of commodities involves a process of deadening, symbolically and materially. Symbolically, it involves deadening in helping to mask the forms of life and their interrelations that make the commodity possible – human labor, for example. Materially, it involves deadening in that commodification necessitates the appropriation of these life forms so that they meet the needs of the commodity production process. Sometimes, this entails the literal killing of something – the cutting down of a tree, for example. At others, it might involve only a partial process of destruction – e.g., the severing of a tree's branches to an extent that the tree's life is shortened.

In the case of ducks and humans at HVFG, we see both types of deadening at play. Until recently, HVFG talked about its practices as "humane." Putting aside the anthropocentric nature of the adjective – which, at its best, would seem to suggest a level of maltreatment of ducks deemed as ethical by human standards, thus not including the voice of the ducks – it is important to note that the culmination of this alleged humaneness entails the duck's slaughter. But right from the beginning, the duck's existence falls far short of its potential (recalling here Galtung's notion of violence).

HVFG 'raises' Mulard ducks, the sterile offspring of male Muscovy ducks and female Pekins, who must be artificially inseminated in order to breed each new generation of duck. Indeed, the name 'Mulard' comes from 'mule,' indicating that the breed can only exist for one generation at a time and requires constant regeneration artificially. Mulard ducklings are also unable to chirp as their parents did as ducklings, and, because they are never allowed to reach maturity, their lifespans are unknown (though Pekin and Muscovy ducks both live past ten or fifteen years) (Wright 2008). As documented by numerous sources, prior to force-feeding, thousands of ducklings are crowded into huge, dry, warehouse-like sheds until they are ready to be force-fed at nine to twelve weeks old (see Tepper 2013). Then, as many as one dozen birds are confined to a wire pen measuring four feet by six feet and force-fed three times per day for the final three weeks of their lives. Holding ducks between their legs, 'feeders' insert a funneled tube down the birds' esophagi and into the ducks' stomachs as they fill the feeding funnel with corn mash. (While HVFG asserts that ducks feel no pain and are physiologically suited for such treatment, research shows that, prior to force-feeding, ducks and geese involved in foie gras production display avoidance behavior, thus indicating aversion for both the person who feeds them and the feeding procedure (European Union 1998).) After the gavage period, the ducks are brought in cages to the slaughterhouse where workers plunge them into electrified water to render them unconscious. The ducks then are bled and de-feathered, their heads, feet, and wings removed, sold, and processed into pet food. The carcasses, minus their appendages, are refrigerated overnight and moved to the next processing phase in which the bodies are dismembered and the necks – that organ so critical to the creation of foie gras – discarded. Finally, the livers are extracted, packaged, and sold.

Despite such treatment, HVFG emphasizes the positive – how, for example, the ducks are less vulnerable than their parents to disease because of their cross-breeding, and how this makes it "feasible for the entire production to take place under one roof." At the same time, the farm claims that the ducks are raised "cage free." While technically true, the birds – waterfowl who never experience life in an aqueous, outdoor environment – live together during the three-week gavage period in groups of ten to twelve inside a roughly four by six foot pen. This artificial, highly contained, and often dark environment exists in contrast to the bucolic imagery displayed in a video on the HVFG website and conjured by a business with the name 'Hudson Valley' (Hudson Valley Foie Gras 2014). According to an investigation by PETA (People for the Ethical Treatment of Animals), HVFG admits to a mortality rate of about 4 percent of its ducks, meaning that forty-one ducks on average die prematurely (before they reach the point of slaughter) each day (People for the Ethical Treatment of Animals n.d.).

As for the immigrant workers from Mexico who do almost all the manual labor at HVFG, they are paid the New York State minimum wage (this despite the fact that individual duck livers are sometimes sold for $80 (Henley 2011, personal communication, October 28)), which was raised to $8 per hour on the last day of 2013. What exacerbates their poor wages is that New York State law categorizes them as 'farmworkers.' As such, they are excluded from legislation that guarantees a day of rest each week, medical benefits, a right to organize, collective bargaining rights, and overtime pay (Herbert 2009; see also Perea 2011). Thus the 'farmworker' classification strips these laborers of the bundle of rights and benefits that accrue to humans, effectively casting foie gras laborers as 'less than human' – just as classification of ducks as nonhuman makes them less worthy than human beings.

During the 1930s, the Roosevelt administration established nationwide rights for all workers except agricultural laborers and domestic workers, beginning with the National Industrial Recovery Act of 1933 and continuing with the National Labor Relations Act of 1935, the 1935 Social Security Act, the Works Progress Administration, and the Fair Labor Standards Act of 1938. New York affirmed these exclusions with the 1937 Labor Relations Act (in which farm-workers were not considered 'employees') and the 1938 state constitution (Gray 2013).

It is no coincidence that labor in the meat and poultry industry, according to the U.S. Government Accountability Office (2005), continues to be one of the most hazardous jobs in the country. Indeed, the bodily costs of production are borne (though to varying degrees) by human and nonhuman alike. The marginal-ized workforce at animal-for-food operations experience 'common' injuries such as "cuts, strains, cumulative trauma, and injuries sustained from falls," not to mention more serious injuries (GAO 2005: 4). According to Christopher Cook, some segments of industrial animal-for-food production have injury rates up to thirty-seven times the average for other industries. In this way, he writes, the work of nonhuman animal slaughter has not changed much since the days of Upton Sinclair's *The Jungle*, except that now the work is done by a largely undocumented immigrant workforce in rural areas of the nation, far from the sterilized aisles of supermarkets or fancy tables of upscale restaurants (Cook 2010: 234).

Human workers at HVFG experience many of the same hazards encountered by the ducks. The entirety of the 'farming' process occurs indoors, in climate-controlled environments that range from hot and humid in the ducklings' barn, to air temperature in the gavage barns, to cold in the slaughterhouse. This manip-ulation of environment and nonhuman lives presents health risks for those engaged in labor: exposure to concentrated amounts of duck urine and feces, slippery floors, and humid air with particulate matter, among others, the latter contributing to the respiratory problems experienced by many workers at the foie gras facility (Margolies 2001).

Sleep deprivation, because of the long hours and the seven-day-a-week schedule, is another hazard. One newspaper account described the life of an *engordador* – a fattener, someone who force-feeds the ducks – at HVFG as follows:

Many nights she sleeps only four hours, from 1:30 to 5:30 a.m., because the schedule often requires her to feed the ducks from 10 p.m. to 1 a.m., then 6 to 9 a.m., and again in the afternoon—more than 1,000 feedings a day.

The worker, identified as Ms. Gonzales, twenty-nine years of age and the mother of two young boys, works sixty-three hours over a seven-day week, meaning nine hours a day – in three-hour segments. The worker elaborated (in Spanish): "Sometimes you get so tired that you fall asleep right in the middle of a feeding. Usually one of the ducks will wake you back up because it's hungry." At the time the article was written, she reported working thirty days straight without a break, saying, "I'd like to have more time to spend with my children. And there's never time to go to church." In between thirty-day cycles (which are now twenty-one-day ones), there are sometimes one or two days before the next cycle starts, but during these times workers often have to clean the barn or unload trucks carrying new ducklings (Greenhouse 2001).

HVFG's owners justify the arduous work schedule by asserting that a day of rest for their workers would reduce the quantity and quality of the final product (Greenhouse 2001). They maintain that the ducks get used to a particular feeder and thus the same person must feed them over the course of the full gavage period. Otherwise, HVFG asserts, the ducks would become frightened and stressed, thus lowering the quality of the final product (Greenhouse 2001). In any case, Izzy Yanay explains, "This notion that [the workers] need to rest is completely futile. They don't like to rest. They want to work seven days" (quoted in Herbert 2009). Putting aside the question of the veracity of Yanay's statement, among what is interesting here is that the socially constructed ducks – ducks constructed as commodities and commodity producers – effectively 'discipline' the workers in that the 'needs' of the ducks determine how and when the *engordadores* labor.

In theory, farmworkers are 'free' to leave – unlike the ducks – if they do not like the conditions at HVFG. To understand why they do not do so requires that we situate the commodification in a larger constellation of social, historical, and geographical relations. At the same time, matters of 'difference' inform how differently marked bodies – of both humans and ducks – experience HVFG and their related (im)mobilities. These are issues we consider in the next section.

Messy intersections: (im)mobility, race, class, gender, and nation at HVFG

As a marginalized, typically unorganized workforce, Latin American farmworkers are subjected to difficult (and dangerous) working conditions for little reward. According to the 1998 National Agricultural Workers Survey, 77 percent of U.S. agricultural workers were born in Mexico, and only 7 percent were U.S.-born non-Latino whites (Johnson and Wiggins 2002: 6–7). This pattern is replicated in eastern New York, including the Hudson Valley, where, according to one study, 90 percent of farmworkers are neither citizens nor legal residents (Gray 2013). Anecdotal evidence suggests that HVFG's non-administrative workforce mirrors this pattern.

For this reason, as well as others, an assertion that workers at HVFG are free to come and go misses the very complicated web of socio-geographic and political-economic relations that shape and circumscribe the choices available to Mexican migrants; it also ignores how in the case of undocumented workers their very 'illegal' status serves to limit their socio-spatial mobility and thus the options they have and the choices they can make (see Nevins 2008 and 2010). At the same time, that most of them are constructed as 'illegal' speaks to the great difficulties Mexican nationals – like non-privileged citizens of countries in the Global South broadly – have in securing authorization to migrate to the United States. As such, if they want to migrate to the United States, they are effectively forced to do so via extralegal means. This is a situation very unlike that which is encountered by, say, Israeli nationals such as Izzy Yanay.

Such differences in mobility are a manifestation of a global context that a number of scholars have characterized as one of global apartheid (e.g., Dalby 1999; Sharma 2006; Nevins 2008 and 2010), one reason being that countries regulate mobility and residence on, among other factors, the basis of geographic origins – one of the foundations of supposed racial distinctions. In doing so, nation-states limit the rights and protections afforded to migrants because of an essentialized characteristic over which migrants have no control. Nation-states also reproduce a world order in which resources (broadly defined), and the right to have access to those resources (through mobility, residence, and employment) are unequally distributed. Thus, it is not simply a coincidence that the owners and managers at HVFG are white, while the workers are Latino and nonwhite. This is not to suggest that such an outcome was inevitable at HVFG, but rather that it is not surprising given the great extent to which race, class, and nation and the regulation of international mobility overlap.

The racial hierarchy of HVFG's workforce is further reproduced by the exclusion of agricultural labor from many of the legal protections and legally guaranteed rights discussed earlier. The Roosevelt administration excluded agricultural and domestic workers in an effort to gain the support of southern Democrats – at a time when these workers, particularly in the U.S. south, were disproportionately African American – who desired to maintain the region's formal racial hierarchy (Perea 2011; Gray 2013). In other words, New York State labor law is based upon, and thus reproduces, a racially segmented labor regime and, with it, its attendant injustices. If we understand race to be about, among other things, the unjust allocation of resources and rights on the basis of racialized difference, and whiteness as embodying a position of privilege within a hierarchy in which blackness embodies subordination, New York State labor law effectively defines non-agricultural and non-domestic workers as white, while effectively (re)producing agricultural and domestic workers as black and thus worthy of fewer rights.

While race and nation inform mobility and the conditions of work at HVFG, so, too, does gender. The official tour of HVFG begins in the office building, staffed entirely by apparently white individuals, who are overwhelmingly female (except for those who appeared to be managers, all of whom are men). In the force-feeding barn, most, if not all, of the feeders are women, who, given their central role in the production process, live on the farm. Owing to the relentless feeding schedule, the *engordadores* spend much of their time on the premises of HVFG, all of it inside the feeding barn. In contrast, male workers engaged in raising the ducks are either in the barn with the ducklings or moving throughout the grounds, apparently attending to the needs of people and nonhuman animals as well as building-maintenance issues. Women engaged in administration are less mobile than their male counterparts (managers) who traverse the farm property and are engaged in all aspects of production. Women – whether attending to desks or ducks – experience spatial stasis to a degree not experienced by male laborers.

The gendered division and placement of bodies within the farm was also apparent in the final destination of the tour, the slaughterhouse. All of the slaughterers are male. Each room has different functions related to the butchering of the ducks and resembles the gendered divisions of labor present in other areas of the farm. While men are exclusively those in charge of the slaughter and initial disassembly, women are in charge of 'preparing' for sale the feather-, appendage-, and bloodless ducks. By this final step the bodies of ducks resemble those one would find at a store or restaurant, and the transformation from living beings into food is nearly complete.

This gendered labor division at the farm constitutes a type of 'toxic mimic' of the typical Western domestic division of labor: women engaged in the (forced) feeding of those charged to their care and the preparation of food (Jensen 2006: 164). On the whole, women are involved (in a perverted way) in the feeding and raising of ducks, and otherwise tied to their domestic environments, which also happen to be on the farm property, whereas men are mobile and conduct the actual slaughter.

The ducks themselves, meanwhile, exist as gendered and sexualized beings in the space of the farm. HVFG, like the other U.S. producers, raises Mulard ducks specifically for foie gras. Mulards have become the breed of choice because of their propensity to grow larger livers in shorter periods of time. However, this ability is more enhanced in male Mulards than females and is why females, upon arrival at HVFG, are separated and then sold to buyers in Trinidad where they are raised for their flesh. Like male calves in the context of the dairy industry and male chicks in the egg industry – both of which involve manipulation of components of female reproductive systems – female ducks in their entirety are useless within the context of foie gras production given the goal of constructing a valuable commodity, one predicated on well-fattened duck livers.

This discrimination relates to the mobilities of male and female Mulards. Both experience simultaneously highly restricted and high levels of transnational mobility. The ducks are born in Canada, shipped to New York and, in the case of the females, sent to the Caribbean. In the case of the males, their end products are literally shipped around the world. In between these strongly regulated 'events' of mobility, the ducks' movement across space, like that of the Mexican laborers who work at HVFG, is very much constrained. This flows from, and highlights the importance of, different spatialities and the power relations embedded within and between them – a matter we consider in the conclusion.

Conclusion: spatial stories

Appreciating matters of intersectionality as they dynamically play out at HVFG helps to illuminate place as more than the collection of physical structures at a location in Sullivan County, New York. As Doreen Massey argues, we should see places as progressive, as flowing over the boundaries of any particular space, time, or society; in other words, we should see places as dynamic or ever-changing, as unbounded in that they shape and are shaped by other places and forces from without across different periods of time, and have multiple identities (Massey 1993, 2005). Acknowledging this allows us to go beyond focusing on places as they *are* currently perceived to exploring *how* they have come into being and thus how they *could be* different in the future.

HVFG, like the United States, is both closed off and open, its boundaries necessarily porous so as to allow for in- and outflows, while also restricting mobility in terms of who and what enters and leaves. As Marcus Henley, HVFG's operations manager told one of us (John) on October 28, 2011 during a tour of HVFG, the farm's best defense against its critics is to have 'glass walls' and that, despite the controversy surrounding foie gras production, they have opened their gates and are probably the most visited poultry farm in the country. Its guestbook, meanwhile, reveals to whom the farm is typically open: restaura-teurs, chefs, visitors from abroad, and those presumably interested ultimately in the *products* of the farm, not the nonhuman and human animal participants. Indeed, John's continued efforts to revisit the farm after his initial tour were rebuffed, each time for a different reason.

The regulated nature of HVFG as a space helps the company tell a story, one that emphasizes certain types of global connections – and narrates them in particular ways – while de-emphasizing or obscuring others, among them the connections that flow from and help to produce violence. This is especially the case of the immigrant workers, who receive far less attention from advocates and critics than do the ducks. As such, the workers are more often than not invisible in the larger discussions surrounding HVFG and foie gras production more broadly. As in agribusiness broadly, these workers "remain a hidden underpinning of the system that brings us the food we enjoy without ever appearing on food labels." In this way, "farmworker invisibility is not unlike the way U.S. consumers are sheltered from reminders of how animals are slaughtered and brought to market in packages or how pesticides are applied to the foods we eat" (Johnson and Wiggins 2002: 8–9).

In the end, if we conceive of HVFG as not just a specific, bounded site, but rather as 'open,' as made up of a network of social relations that transcend the literal space of production, and if those social relations include the consideration of nonhuman actors as serious participants – in other words, if we tell a different story than that usually told – we allow for, and create, new tools of analysis. We also allow for imagining more equitable power geometries than those that ensure the constrained movement and labor of socially disadvantaged groups for the global elite, and that reinforce violent ideologies like speciesism. Noting HVFG's openness highlights its propensity to change through space and time, and articulating what the 'space' of HVFG means by linking it with violent practices of commodification is one way to do so. Instead of HVFG being simply a delimited space in which conditions are undesirable for humans and ducks, one produced by, and productive of, processes of violence and commodification, it is also one of possibility.

Clearly, the spatial dimensions of the intersections between gender, race, nation, mobility, and species at the farm are deep and complicated. The project of identifying and dismantling the related structures of power and injustice, as feminist philosophers Greta Gaard and Lori Gruen (1993: 29) argue, requires that we appreciate "all of the various forms of oppression as central to an understanding of particular institutions." Such a perspective is an essential part of a critical animal geography that seeks to include multiple perspectives, realities, and spaces as part of a project that champions justice, peace, and life for all animals – nonhuman and human.

Another essential component is, as Collard (2014: 154) suggests in discussing that which constitutes being a nonhuman animal of the 'wild' sort (in a normative sense), is autonomy (in this case, from humans). To the extent that the commodification process in relation to HVFG denies or limits such autonomy – by not allowing ducks to feed themselves, for example, and reproducing the socioeconomic and political-geographic marginalization of the workers – it involves the gradual deadening of both. Another way of thinking about this is that autonomy is an essential element of life itself – for humans and nonhumans alike. That is, life of the 'full' sort, one in which animals are allowed to reach their potential unencumbered by socially constructed forms of violence, requires a multifaceted, far-reaching freedom. However, this is not to suggest a freedom that is unencumbered because the very fact of interactions with others and the world around us necessitates limits. But these limits should be ones that are established through negotiation, through what Doreen Massey (2005: 181) calls a "politics of connectivity" – one that requires respect for all those involved. Hence, true respect for humans and ducks at HVFG requires not only the dismantling of the company as it now stands and of the larger political-economic and socio-spatial contexts that flow from and enable it, but the creation of dynamic, new configurations entailing a politics that is radically democratic in a way that is anti-speciesist.

Acknowledgments

We wish to thank Rosemary Collard, Greta Gaard, Kathryn Gillespie, and Keith Lindner for valuable feedback on an earlier version of the manuscript as well as Kaitlin Reed for editorial and research assistance and feedback on various versions of the manuscript.

References

Appadurai, A 1986, *The social life of things: Commodities in cultural perspective*, Cambridge University Press, Cambridge.
Collard, R-C 2014, 'Putting animals back together, taking commodities apart,' *The Annals of the Association of American Geographers*, vol. 104 (1), pp. 151–165.
Collard, R-C and Dempsey, J 2013, 'Life for sale? The politics of lively commodities,' *Environment and Planning A*, vol. 45 (11), pp. 2682–2699.
Cook, CD 2010, 'Sliced and diced: The labor you eat,' in D Imhoff (ed.), *The CAFO reader: The tragedy of industrial animal factories*, University of California Press, Berkeley.
Cresswell, T 2006, *On the move: Mobility in the modern western world*, Taylor & Francis, New York.
Dalby, S 1999, 'Globalisation or global apartheid? Boundaries and knowledge in postmodern times,' in D Newman (ed.), *Boundaries, territory, and postmodernity*, Frank Cass, London.
Deckha, M 2012, 'Toward a postcolonial, posthumanist feminist theory: Centralizing race and culture in feminist work on nonhuman animals,' *Hypatia*, vol. 27 (3), pp. 527–545.
Dunayer, J 2004, *Speciesism*, Ryce Publishing, Derwood, MD.
Eatocracy 2012, 'Clarified – is foie gras fair game or foul play?,' June 13. Viewed January 17, 2014. http://eatocracy.cnn.com/2012/06/13/clarified-is-foie-gras-fair-game-or-foul-play/.
European Union Scientific Committee on Animal Health and Animal Welfare 1998, 'Welfare aspects of the production of foie gras in ducks and geese,' December. Viewed January 21, 2014. http://ec.europa.eu/food/animal/welfare/international/out17_en.pdf.

Gaard, G 2002, 'Vegetarian ecofeminism: A review essay,' *Frontiers*, vol. 23 (3), pp. 117–146.

Gaard, G and Gruen, L 1993, 'Ecofeminism: Toward global justice and planetary health,' *Society and Nature*, vol. 2 (1), pp. 1–35.

Galtung, J 1969, 'Violence, peace, and peace research,' *Journal of Peace Research*, vol. 6 (3), pp. 167–191.

Galtung, J 1990, 'Cultural violence,' *Journal of Peace Research*, vol. 27 (3), pp. 291–305.

Glassman, J 2006, 'Primitive accumulation, accumulation by dispossession, accumulation by "extra-economic" means,' *Progress in Human Geography*, vol. 30 (5), pp. 608–625.

Gray, M 2013, *Labor and the locavore: The making of a comprehensive food ethic*, University of California Press, Berkeley.

Greenhouse, S 2001, 'No days off at foie gras farm; workers complain, but owner cites stress on ducks,' *New York Times*, April 2. Viewed January 20, 2014. www.nytimes.com/2001/04/02/nyregion/no-days-off-at-foie-gras-farm-workers-complain-but-owner-cites-stress-on-ducks.html?scp=5&sq=hudson+valley+foie+gras&st=cse&pagewanted=all.

Gruen, L 2001, *Ethics and animals: An introduction*, Cambridge University Press, Cambridge.

Guillermo, K 2013, 'Ducks have law – and right – on their side,' *The Huffington Post*, December 4. Viewed December 7, 2014. www.huffingtonpost.com/kathy-guillermo/ducks-have-lawand-righton_b_3861097.html.

Harvey, D 2003, *The new imperialism*, Oxford University Press, New York.

Herbert, B 2009, 'State of shame,' *New York Times*, June 9, p. A27. Viewed September 15, 2014. www.nytimes.com/2009/06/09/opinion/09herbert.html.

Hudson Valley Foie Gras 2014, 'About Hudson Valley Foie Gras'. Viewed December 20, 2013. www.hudsonvalleyfoiegras.com/index.php/about-hvfg.

Jensen, D 2006, *Endgame*, Seven Stories Press, New York.

Johnson Jr, CD and Wiggins, M (eds) 2002, *The human cost of food: Farmworkers' lives, labor, and advocacy*, pp. 6–7, University of Texas Press, Austin.

Margolies, K 2001, *Training needs assessment of farm workers in Orange and Sullivan Counties, NY*. Retrieved from Cornell University, ILR School site. Viewed September 22, 2014. http://digitalcommons.ilr.cornell.edu/articles/176.

Marx, K 1992 [1867], *Capital*, Vol. 1, International Publishers, New York.

Massey, D 1993, 'Power-geometry and a progressive sense of place,' in J Bird, B Curtis, T Putnam and L Tickner (eds), *Mapping the futures: Local cultures, global change*, Routledge, London.

Massey, D 2005, *For space*, Sage, Thousand Oaks, CA.

Nevins, J 2008, *Dying to live: A story of U.S. immigration in an age of global apartheid*, Open Media/City Lights Books, San Francisco.

Nevins, J 2010, *Operation Gatekeeper and beyond: The war on 'illegals' and the remaking of the U.S.-Mexico boundary*, Routledge, New York.

Nevins, J and Peluso, N 2008, 'Introduction: commoditization in Southeast Asia,' in J Nevins and N Peluso (eds), *Taking southeast Asia to market: Commodities, nature, and people in the neoliberal age*, pp. 1–25, Cornell University Press, Ithaca.

People for the Ethical Treatment of Animals n.d., 'Foie gras cruelty exposed'. Viewed February 3, 2014. https://secure.peta.org/site/Advocacy?cmd=display&page=UserAction&id=4803.

Perea, JF 2011, 'The echoes of slavery: Recognizing the racist origins of the agricultural and domestic worker exclusion from the National Labor Relations Act,' *Ohio State Law Journal*, vol. 72 (1), pp. 95–138.

De Rilke, V 2008, *Duck*, Reaktion Books, London.

Sella, M 2005, 'Does a duck have a soul?,' *New York Magazine*, June 27. Viewed September 15, 2014. http://nymag.com/nymetro/food/features/12071/.

Sharma, N 2006, *Home economics: Nationalism and the making of 'migrant workers' in Canada*, University of Toronto Press, Toronto.

Tepper, R 2012, 'Undercover foie gras footage shot at Hudson Valley Foie Gras alleges cruel practices,' *The Huffington Post*, June 12. Viewed January 17, 2014. www.huffingtonpost.com/2013/06/12/undercover-foie-gras-video-hudson-valley_n_3429492.html.

U.S. Government Accountability Office 2005, 'Safety in the meat and poultry industry, while improving, could be further strengthened,' a report to the Ranking Minority Member, Committee on Health, Education, Labor and Pensions, U.S. Senate, January. Viewed February 3, 2014. www.gao.gov/new.items/d0596.pdf.

Wright, L 2008, *Choosing and keeping ducks and geese: A beginner's guide to identification, care, and husbandry of over 35 species*, TFH Publications, Neptune City, NJ.

7 Species, race, and culture in the space of wildlife management

Anastasia Yarbrough

Introduction

Wildlife management and conservation have long been considered the domain of upper-middle-class white privilege in the United States (Harry *et al.* 1969; Fox 1981; Warren 1997; Taylor 1997, 2002). This chapter explores how racism and colonialism after two centuries continue to play a role in modern wildlife management in the United States. In particular, I focus on the extent to which the historical American conservation movement has used racism and colonialism as weapons to manage wild animals as resources for the primary enjoyment of upper-middle-class whites while simultaneously marginalizing people of color (e.g., African Americans, Native Americans, and Latinos) and excluding these groups from full participation in wildlife management. I also examine to a brief extent how conventional wildlife management practices deny animal agency and how population management as a paradigm affects the ways in which people of color and wildlife are 'managed' in the United States. Finally, I consider the opportunities and pitfalls of using an environmental justice framework as an alternative to conservation for wildlife advocacy and community empowerment. In alignment with critical animal geographies, I attempt to decentralize the status quo of wildlife management and reimagine wildlife management in a way that genuinely includes wild animals and people of color as legitimate subjects in a community of biocultural diversity.

Before I proceed, I would like to clarify my use of language. Often, wildlife is used as a singular noun to summarize a large, diverse group of animals. Similar to the way that a species is referenced in the singular (i.e., the Virginia Bat), wildlife is more a generalized idea than an accurate depiction of those beings we refer to as 'wild animals.' However, to minimize confusion, in this chapter, I will use the term 'wildlife' interchangeably with 'wild animals.' When I say wild animals, I am signifying those nonhuman animals who currently and have for multiple generations lived autonomously from human technology. This definition, as I use it, also includes animals we label as 'feral' or 're-wilded.' Animals who are wild tend to exhibit certain qualities such as self-reliance, self-determination, and reproductive freedom. Even these qualities are being contested as humans breed wild animals in captivity and urban, agricultural, and energy development make self-determination life-threatening for some animals. At the moment, 'wild animals' is the most precise phrase I know to capture this state of being. Therefore, in the likeness of Jennifer Wolch (1998), I will refer to wild animals consistently throughout this chapter to differentiate from captives such as pets, farmed animals, animals in laboratories, and animals in zoos and aquaria.

The need for wildlife management: from past to present

When Europeans began to colonize what later became the United States, early Anglo-American male authors reported the abundance of wild animals roaming free in close proximity to, and often through, European settlements. They "found wildlife everywhere because they traveled and settled where it was most concentrated" (Trefethen 1966: 18). Europeans also changed the landscape everywhere they went, namely through deforestation, and thereby created habitat attractive to, and supportive of, species of economic value (Tober 1981). Thus, a massive political economy surrounding wild animals' bodies in the United States grew. Through the 1600s to the mid-1800s, economic opportunities abounded for white men in the skin, fur, plume, and bone trades so much that many laboring men dropped their low-paying jobs in exchange for high-paying adventures of the professional hunter (Tober 1981). After the Civil War, commercially valuable animals became harder and harder to find. As white outdoorsmen began to voice concern over this 'loss,' the birth of a new movement began to take off in the United States.

The need for wildlife management arose out of the use of wild animal bodies and wild spaces. Hunting restrictions and predator control were the earliest responses to manage the dynamics of wild animal populations, and they remain staple activities of wildlife management agencies nationwide. By the turn of the twentieth century, wildlife conservation emerged out of this hunting and control-based paradigm (e.g., Hornday 1913). The recreational activity often labeled sports hunting was a powerful catalyst in shaping wildlife management. Affluent sports hunters like Aldo Leopold and JN 'Ding' Darling (founder of the National Wildlife Federation) influenced federal conservation policy that still stands today (such as the Pittman–Robertson Act and the establishment of wildlife refuges) as wildlife management became a formal institution comparable to agriculture. (Note the early refuges did not offer wild animals refuge from hunting and angling and that remains a feature of U.S. Fish & Wildlife Service's management of wildlife refuges.) Managers set regulations to the 'harvest' and the use of animals' bodies in order to balance hunting losses, natural losses, and production to ensure healthy populations will be forever available for recreational killing (Hernandez and Guthery 2012). Aldo Leopold – a man considered the father of U.S. American wildlife science and management – captures the institution perfectly:

> The game manager manipulates animals and vegetation to produce a game crop.... Herein lies the social significance of game management. It promulgates no doctrine, it simply asks for land and the chance to show that farm,

forest, and wild life products can be grown on it, to the mutual advantage of each other, of the landowner, and of the public.

(Leopold 1933: 420, 422)

Yet, wildlife management was and is fundamentally ideological, its evolution dominated by the ideas and desires of affluent white men. The commercialization and production of wild animals, the idea that by owning land one owns the wild inhabitants on it, the consumption of wildlife as a crop and renewable resource, the importance of empty or scarcely populated land to 'grow wildlife' in the interests of the owner – all have been prominent ideological tools in shaping wildlife management. They by no means originated with Leopold. Throughout European history, affluent landowners either directly passed laws or influenced the passing of laws to protect elite hunting interests (Gilbert and Dobbs 2001). William the Conqueror was the first to pass the Royal Forest Law in England which imposed harsh penalties on peasants and non-royal persons who attempted to hunt animals in the private royal forest, which was the majority of forest in the British Isles.

U.S. American values for wildlife diverged from Europe only as white men began to prize wild animals as indicators of equal opportunity and access (ironically, this equal opportunity and access did not extend to white women or people of color). For white men, wild animals were 'there for the taking.' They were "a common heritage, not subject to restrictive controls which smacked of Old World class structure" (Tober 1981: 12). Wild animals were an opportunity for white men to do what they wanted, a luxury most colonial migrants could not enjoy in their previous countries. Wild animals were a bounty of resources in a new world that white men could make in their image.

In 1930, Leopold and a committee of fourteen at the American Game Conference drafted what are now considered the basic policy requirements for public wildlife management (Gilbert and Dobbs 2001). From this foundation, wildlife management agencies nationwide share these criteria in one form or another:

1   *Managed populations must meet protection, food, and cover requirements.* In order to sustain wild animal populations at the desired level, managers must first begin with the basic requirements for individual and community survival.
2   *Landowners get incentives to participate in conservation programs.* This continues to be important because the vast majority of land in the United States is privately owned.
3   *Game animals are classified into farm, forest and range, and wilderness.* Game animals were those animals whose bodies were considered of commercial or aesthetic value, and they were classified in habitats where they were most common or easiest 'to grow.' These initial habitats were not scientific or ecological but rather based on the economic uses or valuation of land at the time. These categories continue to be used in public wildlife, forest, and land management today.

4   *Wildlife management needs the continual support of science, funding, and public-sportsman cooperation.* This is perhaps the most crucial aspect of wildlife management for it sets the values behind wildlife management and thus the direction of policymaking, stakeholder engagement, and research questions explored.

By 1937, the Pittman–Robertson Act was passed, allowing federal and state taxation on arms and ammunition to fund wildlife scientific research that in turn informs state-level and federal wildlife management. In fact, the tax continues to fund over 75 percent of "approved wildlife projects," according to the U.S. Fish & Wildlife Service, extending to the taxation of not just guns and bullets, but also bows, arrows, and other killing equipment used by sports hunters. It remains one of the most important pieces of legislation to wildlife management as an institution, and it is a reinforcing justification for hunters and gun enthusiasts to strong-arm wildlife management decisions.

The last tenet of public wildlife management mentioned above is the one most relevant to this chapter. Wildlife management agencies, in their heavy reliance on sports hunter and angler cooperation, are structured in such a way that the primary stakeholders who influence what happens to wild animals are those who *use* wild animals as economic *and* aesthetic resources, primarily for the purposes of trapping and killing. Thus, the usefulness of animals' bodies and the pleasure of killing go hand in hand. This standard in wildlife management and conservation shapes the image of wildlife management in the public imagination. In my educational experience with university wildlife biology programs, the course curricula extend conservation principles and population management applications to wild animals in general, but not without stressing the importance of appeasing sports hunters and land owners.

This ideology surrounding wildlife management has a history in the United States that has roots in the historical and contemporary identities of white masculinity. In fact, one could argue in depth that wildlife management runs parallel to the conservation of particular white male hero identities. To some, regulating the harvest of wild animal bodies is a biblically ordained duty enforced by God (Gilbert and Dobbs 2001). Early European colonizers saw the numerous wild animals, unafraid of human contact, as bountiful resources from God as well as objects through which men could prove themselves among their peers. The more 'scarce' wild animals became, the greater the drive for a man to prove himself (Tober 1981). Theodore Roosevelt goes into further depth on wild animals as devices for exercising violence, acts he perceived as necessary to advance the appropriate idea of masculinity:

> In hunting, the finding and killing of the game is after all but a part of the whole.... The chase is among the best of all national pastimes; it cultivates that vigorous manliness for the lack of which in a nation, as in an individual, the possession of no other qualities can possibly atone.
>
> (Roosevelt 1893: 11)

While he did not name his ideas as exclusively the rite of passage for white men, policies during and after his administration say otherwise. African Americans and Latinos were not allowed access to public lands, and Native peoples –considered hostiles and wards of the United States – were continually forced out of what became federally managed public lands and onto reservations. Roosevelt supported the Yellowstone Forest Reserve (the presiding agency over Yellowstone at the time) and the discouragement of Native people's occupancy of Yellowstone along with the active killing of cougars, wolves, and bears in Yellowstone. He would assist personally with executions of cougars and wolves, referring to both groups as "beasts of desolation and waste" and "bloodthirsty," "cowardly" predators with "a desire for bloodshed which they lack the courage to realize" (Roosevelt 1927: 305). And, throughout his presidency, he would hold fast to his antipathetic views toward Native peoples: "I don't go so far as to think that the only good Indians are dead Indians, but I believe nine out of ten are, and I shouldn't like to inquire too closely into the case of the tenth" (Dyer 1992: 86). As the political economy of wild animal bodies rose, the cultural aesthetics and scientific study surrounding wildlife grew. Initiated by urban, elite, white male visionaries from the northeast, the U.S. American preservation movement facilitated the rise in production of natural history literature and championed a romanticized cultural aesthetic of wild animals and 'wilderness' as beautiful, spiritually divine sanctuaries from hectic, decaying civil life (Tober 1981). Men like George Catlin and Henry David Thoreau were not hunters or anglers, but like the hunters of the conservation movement, they relied on wild animals and wilderness as symbols of their masculinity, humanity, and U.S. American identity (Thoreau 1854; Spence 2000). However, unlike prominent hunters of their time, they championed passive consumptive values, almost nonconsumptive values, of wild animals.

Wilderness was crucial to the practice of habitat conservation and the multilayered valuing of wild animals in human lives. It also played a role in forming the wildlife public trust, the idea that the public collectively owns wild animals, as opposed to selective ownership by individual private landowners or no ownership at all. While this idea did not become part of United States environmental policy, it remains a concept supported by scholars and scientists in wildlife management (Organ and Batcheller 2008; Jacobson *et al.* 2010; Bruskotter *et al.* 2011).

Wild animals occupy and belong to wilderness. Wilderness contains wildlife. The name 'wilderness' comes from the old English word for 'the realm of the animals.' Until the formation of 'Indian reservations' in the mid-1800s, U.S. American romantic art and literature (Fenimore Cooper 1827, George Catlin's Indian Gallery) often depicted Native peoples as belonging to wilderness. However, white public attitudes about wilderness and the place of people in it began to shift as attitudes toward Native peoples grew less romantic. With the installation of national parks and public wilderness areas, Native peoples came to be seen as nuisances rather than natural heritage icons (Spence 2000). Native communities across the country were discouraged or forcefully removed from national parks over the greater half of the twentieth century. Expressed through the Wilderness Act of 1964, humans became "visitors [to wilderness] who do not remain," and therefore must be managed within the visitor status. Mark David Spence (2000) argues that maintaining the wilderness idea – the idea that humans do not belong in wilderness and that wild animals must forever be contained in wilderness – was crucial to the creation of wilderness. In other words, wilderness serves as unoccupied, empty land where valued wildlife are but animated symbols, and forests the places upon which the Euro-American imagination can be exercised and can rest. Colonialism was the driving force behind this creation (e.g., Guha 1989; Cronon 1995).

The last four decades of research in wildlife ecology and behavior have forced both wildlife management and wild animals to evolve. Animals are extending themselves and exploiting opportunities in the name of survival. Shore birds use stopover habitat at urban water treatment facilities (e.g., Dececco and Cooper 1996), songbirds sing louder to adjust to urban noise (e.g., Slabbekoorn and den Boer-Visser 2008; Bermúdez-Cuamatzin *et al.* 2011), and a multitude of mammals such as raccoons, bats, squirrels, and opossums raise families with the aid of buildings and refuse. With urban/suburban sprawl and increasing development, wildlife habitat in the city is not just an aspiration for urban planning, it is an ever-pressing reality we must face (Wolch 1998). Bird and mammal species often labeled as common are thriving in cities, subsisting on the refuse and aggregation of resources consequent of urban life. Urban/suburban development and international trade facilitate the growth of immigrant wild animal populations, many of whom are federally listed as invasive species (e.g., National Invasive Species Information Center). Animals who were never considered economically or even aesthetically valuable are added to state and federal wildlife agencies' lists of conservation concerns. Biodiversity has become a new way of thinking about management, with biodiversity as a goal rather than the proliferation of select, economically valuable species, but it has not been adopted at federal and state levels of management, and thus remains confined to academic ideals of conservation biology.

Further change is occurring in how Americans 'consume' wild animals. U.S. Fish & Wildlife Service (2011) reports that approximately 71.8 million people aged sixteen and over were engaged in wildlife watching in 2010, 34.4 million more than those who angled and hunted. This trend toward wildlife watching has been increasing in the American populous for over ten years, and it has sparked scholars and researchers to rethink the role, philosophy, and practices of wildlife management (Schuett *et al.* 2008). But even with this shift and the influx of new concerns, new thought, new problems, and new people (such as the human dimensions of wildlife), wildlife agencies continue programs as though sports hunters and their 'harvest' are the main agenda, and sports hunters and anglers remain the primary constituent for wildlife agencies. Just visit the websites of Texas Parks & Wildlife, North Carolina Wildlife Resources Commission, New York State Department of Environmental Conservation, and the Washington Department of Fish & Wildlife, and the first feature you are likely to see is an advertisement for angling or hunting. Stephen Kellert laments this overreliance on sports hunters and anglers in wildlife management:

> Financial dependence on taxing sportsmen ... has distorted the goals of the profession by encouraging the wildlife management field to accommodate itself to the needs of this clientele rather than the condition of wildlife more generally.

> (Kellert 1996: 208)

I suggest that wildlife management's financial crux is not so much a distortion of goals as it is an indication of its heritage. Present day ideology, principles, and goals of wildlife management would not exist without the white sportsmen who owned land. Because wildlife management emerged as an institution constructed by white sportsmen *for* white sportsmen, wild animals' interests never played a key role in the goals of wildlife management. Therefore, it is no surprise that the condition of wildlife more generally beyond game species population dynamics would not be of major concern in U.S. wildlife management. Since the publication of Kellert's *The Value of Nature*, wildlife management has expanded to accommodate the conservation of endangered and threatened species as U.S. American public values toward wild animals shift. However, hunting and angling programs are at the core of wildlife management programs, so the financial dependency on such activities has not changed in the last two decades.

In alignment with the legacy of wildlife management and its initial premise, sportsmen continue to provide the most revenue through gun and permit sales (USFWS 2011). Although more people are participating in wildlife viewing, they are less able to influence decision-making or shape public wildlife programs. This is likely because "wildlife-associated recreation" is more difficult to define, track, and pull revenue from. That has not stopped research in the human dimensions of wildlife from exploring opportunities in wildlife viewing for financial support and policy influence in wildlife management (McCool 2008).

The trade in wild animals' bodies certainly has not disappeared in the name of management and conservation. In fact, international entities such as the World Wildlife Fund, TRAFFIC, and IUCN distinguish "illegal wildlife trade" from "legal wildlife trade," with the latter as an important commercial practice for sustainable economic development. However, while easy enough to distinguish in theory, the 'legal' and 'illegal' trades of wild animals are difficult to differentiate in practice. Illegal wildlife trade is currently the third largest black market economy after guns and drugs (Izzo 2010). This includes not just body parts hot on the market (e.g., elephant ivory and moon bear bile), but also living wild animals. Live wild animal trade is particularly relevant to the United States as it is one of the largest importers of wild animals in the world (Smith *et al.* 2009) and it fuels roadside exotic attractions (Green 1996) and 'canned hunts.'

Commercial hunting in the United States has evolved into what animal rights activists call the canned hunt. At least 330 active (National Hunting Club Listing

2014) private hunting clubs and commercial hunting operations (CHOs) in the U.S. sell the pleasure of killing animals to anyone willing to pay for it. These private hunting clubs cost much more to access than public lands, making them demographically exclusive. Their hunting seasons are unlimited, unlike wildlife management areas or wildlife refuges where they depend on the species in question. And club membership comes with the guarantee that the customer will always walk away with a body. It is no surprise, then, that the majority of hunters in the United States hunt animals on private lands (USFWS 2011).

Wildlife Systems, Inc., a Texas-based CHO claiming over 500,000 acres of private land, advertises its colonial might by offering customers the opportunity to kill animals from native as well as 'exotic' species. I do not go into depth on the multitude of private safari hunting clubs marketing the killing of 'exotic' animals. The international trade of living wild animals extends beyond the scope of this chapter, but the importance of modern wildlife trafficking in shaping institutions such as CHOs, laboratory research facilities, and urban zoological gardens cannot go unmentioned (for more, see Meredith Gore's work in conservation criminology; Green 1996; the Committee on Natural Resources 2008; and the American Anti-Vivisection Society [n.d.] report on rhesus macaques). Another CHO in Texas (Texas Disposal Systems), specializing in the breeding of 'exotics,' advertise proudly on their website that their "wildlife reserve is used primarily for animal production for [their] private collection, as well as endangered species conservation." Nowadays, commercial hunting operates under the guise of endangered species conservation. This makes the conservation framework even less reliable for wild animal advocacy than it already was. I explore this subject again in the last two sections.

Racism and colonialism drives the need for wildlife management

The exclusion of people of color from wildlife management and conservation was not happenstance. It involved an intentional multi-century process shaped and maintained by white affluent men. Before wildlife conservation became an issue, water development and sustained-yield forestry dominated the stage of conservation (Hays 1989). Both issues were placed in western United States where ranching and logging activities were the exclusive domain of white settlers. These activities were not necessarily activities of the elite, but recreational hunting – the harbinger of wildlife conservation – was. It was unthinkable for people of color to have been granted access to private hunting and angling clubs which constituted the early wildlife conservation movement (Fox 1981). When men of color showed interest in these activities, they were denied the opportunity, arrested, or killed. State game laws across the United States were backed by elite white conservationists to exclude ethnic minorities they believed did not uphold the 'proper' hunting ideals (Warren 1997). In Kentucky, for over a century, arbitrary laws such as black people not being allowed to hunt on Sundays, made even subsistence hunting challenging to maintain (Forehand

1996). State fish and wildlife officials enforcing racist game laws opened fire on Native hunters, killing several men, in Wyoming in 1903 and in Montana in 1908, and New Mexico wildlife officials openly encouraged ranchers in the state to kill Native hunters in 1905 (Warren 1997). This violence and discrimination continued under the guise of 'conservation' and 'sound wildlife management.'

Conservationists and wildlife enthusiasts of the time behaved in ways that echoed national attitudes toward African Americans, Native Americans, and other ethnic minorities. The Sierra Club chapters in southern California deliberately excluded African Americans, Native Americans, Jews, Chicanos, and other ethnic minorities from membership for over half a century. Wildlife conservationists' exclusionary practices extended to nationality as well. Early conservationists often believed that Italians, Greeks, and Eastern Europeans were responsible for "depleting wildlife populations through old-world hunting customs" (Jordan and Snow 1992: 77). Fox concludes that racism and nationalism perpetuated throughout wildlife conservation NGOs were but a product of the time:

> Racist by modern standards, their attitudes cannot be separated from historical factors specific to the period: the new immigration, the tensions of modernization and rampant progress, cancerous urbanization complemented by the end of the frontier, and the bunker mentality among the old, displaced patricians. Progressive-era conservationists expressed nativist ideas more as prisoners of their time than as conservationists.
>
> (Fox 1981: 351)

However, racism is not just a thing of the past. Ruland (2012) revealed that the Texas Parks & Wildlife Department (TP&W) has been intentionally discriminating against black people in their hiring practices since the 1970s. Black game wardens found it virtually impossible to be promoted from entry-level positions, and in fact, no black wardens had been hired at all since 2005. Much of this was due to pervasive fear among white wardens that they would have to answer to a black warden in a leadership position. Such a minor shift in leadership as that is enough to shift the status quo of wildlife management, or at least give the impression of a shift. As demonstrated with the TP&W, that simply could not be allowed.

In addition to the intentional exclusion from participation in wildlife-related recreation and employment, people of color faced styles of population management similar to those imposed on 'nuisance' species. Free black people in the post-reconstruction United States were terrorized away from any direction that would foster ecological and economic autonomy. Many freed slaves were forced to return to the plantations on which they were born only to become tenants and sharecroppers. Many black people sought refuge and better economic opportunity north and west but instead faced job and housing discrimination and racial tension with low-income European immigrant workers. Housing authorities in cities like Chicago and Detroit managed vast migrations of southern black people into ghettoes otherwise known as public housing – a management process not very different from the reservation. Both management systems herded people into centrally located yet isolated communities. Both were constructed in desolate habitats with little to no economic opportunity. Both made residents extremely vulnerable to and dependent upon the presiding agency for basic needs, leaving future generations without the knowledge for self-sufficiency or the confidence for long-lasting self-determination.

Meanwhile, wild animal species in North America were removed, displaced, and enslaved in large numbers between 1820 and 1960. European immigrants drove both heath hens and passenger pigeons to extinction. Free-roaming buffalo herds were completely decimated across the country, to where the majority of buffaloes to this day live in captivity on ranches. Between 1880 and 1920, cougars and wolves were executed en masse and removed from public lands across the country. After many years of being killed for their fur (and in retaliation for Euro-Americans' conflict with their cultural-ecological ways), beaver communities were destroyed throughout most of the United States and did not re-establish themselves until U.S. values about beavers shifted from products and pests to ecosystem engineers. White America's management of African Americans, Native Americans, and select wild animal species via displacement, captivity, and persecution were happening at the same time in U.S. history. I show these events parallel because they reflect the population management paradigm particular to the United States' efforts to subjugate and control seemingly inferior races or species, and this paradigm is apparent across governmental agencies. This paradigm that encourages subjugation over cooperation, exclusion over inclusion, continues to produce population management institutions that lack compassion and diversity.

The absence of people of color in wildlife-related activities

Wildlife management has been a predominately white male institution since its inception, although today women over twenty-five are much more active in non-lethal recreation such as wildlife watching and photography (USFWS 2011). White women are more likely to occupy leadership and research positions in wildlife agencies, natural resource programs at universities, and NGOs, but they remain greatly outnumbered by white men (Nicholson *et al.* 2008). White people with annual incomes ranging between $30,000 and $100,000 are the largest consumers in wildlife recreation (USFWS 2011). Whether it's angling, hunting, or simply watching animals be, white people are a majority of over 85 percent. Sports hunting continues to be the exclusive domain of middle- to high-income white men. Wildlife watching was the only activity not so determined by income or gender, most likely because one does not have to travel far from home to watch animals and it does not require many tools and equipment to do it. However, people of color still are not participating in these activities.

Stephen Kellert (1996) suggests that a person's education level makes a difference in influencing that person's level of concern toward animals and the natural world. The higher one's education (i.e., college degree and beyond), the more likely one is to demonstrate greater concern, affection, interest, and knowledge of animals. People with a high-school education or less are more likely to express exploitative, authoritarian attitudes toward animals and the natural world. However, this generalization does not hold with the majority of African Americans. Based on his research, the difference between black and white people is sharpest at the level of college education. Rather than show an increase in concern for wild animals and wild places, college-educated black people reveal values more similar to low-income urban dwellers with a high-school education or less – less appreciation, less recreational interest, and less willingness to support wildlife conservation programs. Kellert concludes that most black people view the experience and conservation of 'nature' as "being of only marginal concern."

Many theories have been postulated over the years as to why Americans of color (with the exception of Native Americans) have demonstrated little interest in wildlife matters. One problem is that the standards set for gauging interest are rooted in European notions of wildlife recreation. As populations of people of color in the United States continue to rise (particularly for Latinos and Asian Americans), fewer people in general are involved in wildlife-related outdoor recreation, but especially people of color. Wildlife recreation requires time, access, good health, commitment, and money. Leonard (2007) reported that time was the most important reason why people quit activities like hunting and angling. Most notably, time constraints were most acute for people of color and people living in large cities (Leonard 2007). The most expensive form of wildlife recreation by far is sports hunting, and it is not surprising that the majority of people of color in the United States are not involved in it. In addition, the culture of wildlife recreation often includes taking trips away from home and work to make extended efforts in finding a particular animal of interest. This exacerbates the luxury and privilege people of color already associate with wildlife recreation.

Many people of color label wildlife-related recreation as culturally irrelevant, since both wildlife recreation and wildlife management have historically excluded them (Floyd 1999). People of color are more likely to fit Kellert's values framework of utilitarian or no appreciation of wildlife because of educational and job disparities people of color on average face (American Psychological Association 2012; Cooper 2014). Additionally, discrimination and harassment remain an issue in public places of outdoor recreation. Just the anticipation of harassment discourages people of color from visiting parks and outdoor recreation areas away from home (Allison 2000). Finally, the racial makeup of employees at park and recreation agencies is still predominately white, many of whom may lack the skill and interest in relating with people of color or awakening to their own racialized consciousness (Jordan and Snow 1992; Floyd 1999).

A major ethical difference between peoples of color and white people is how they place themselves in ecological communities and in the realities of those communities. In both cases of the black and Latino movement to ecologically and economically revitalize the South Bronx, and the Native Alaskan struggle to maintain and enforce treaty rights for traditional hunting and angling practices, the environmental and wildlife concerns include the communities of color directly. They are advocating for the places and ecological contexts in which they live. Wild animals are included in the efforts for environmental justice, but rather than take center stage as the hunting, aesthetic, or ideological objects of concern, they are presented as background objects to the greater environment. They are but a piece of the greater concern that is the human–environment relationship. While this is not an ideal form of including wild animals, it is a more direct way in which the majority of people of color engage wildlife issues. When animal rights and environmental activists protest Native Alaskan seal hunting, the protestors tend not to live in the region of concern. Their relationship with the seals is primarily ideological, aesthetic, and distant. Their ecological and economic lives are generally stable, and the issue of cultural survival is of no concern. Therefore, the issue becomes the seals whom the activists do not have a historical–cultural–ecological relationship with versus the Native Alaskans living on reservations struggling to maintain health, lifeways, and cultural identity. More so than lack of concern for the persistence and survival of wild animal communities, I suggest the cultural spaces of wildlife advocacy, wildlife research, and wildlife management and conservation deter people of color from involvement. And it is the racist and colonial legacy of these spaces that serve as bitter reminders of exploitation, displacement, and exclusion.

It is not surprising that people of color in rural countryside or cities do not have a strong cultural impetus to include wild animals beyond the role of resource. After all, if the paradigm for wildlife protection is based on the use of wild animals (which is the message one receives from a large wildlife organization like the Sierra Club or National Wildlife Federation), and if people of color have been historically denied the use of wild animals to reinforce their cultural identity and 'natural heritage,' then why should they go out of their way to protect wild animals as their 'natural heritage,' animals living half a continent away, when so many communities of color in the United States are still struggling to meet basic needs, address social disparities, and improve their quality of life and environmental conditions? An alternative framework for relating with, living with, and appreciating wild animals is needed.

Animal agency denied

Wildlife management was never developed to serve the interests of wild animals. In fact, wildlife management is often predicated on the ignorance and violation of individual animal agency and their communities. Some recent studies have shown how individual animals and animal communities across species change their ways in response to what happens in the environment. Several studies have been done with birds and whales demonstrating that under noisy environmental conditions individuals consistently raise their voices above the noise (e.g., Patricelli and

Blickley 2006; Slabbekoorn and Ripmeester 2008; Holt *et al.* 2009; Parks *et al.* 2011; Bermúdez-Cuamatzin *et al.* 2011). Migratory birds, such as marsh harriers, demonstrate flexibility in their decision-making such that individuals do not often repeat the same migratory routes twice (Vardanis *et al.* 2011). And male livebearing fishes will use public information and what they know about other males with whom they live in community to inform how they pursue sexual partners (Bierbach *et al.* 2011). Biologists often refer to these behaviors as adaptations or "behavioral plasticity" (which is a scientific way of saying that an individual animal's behavior can change and vary over a single lifetime), and some will use the term 'culture' adding to the extensive debate of "animal cultures" in the scientific community (e.g., Laland and Janik 2006; Whiten and van Schaik 2007; Laland and Galef 2009). By acknowledging the agency of animals, these studies confirm that animals are not fixed automatons. Wild animals are dynamic, living beings who make decisions, change their minds, relocate, and behave in the interests of subjective motivations.

As a result, they do not always conform to what is expected of them. Some species will even have the nerve to 'get in the way.' This often leads to lethal consequences for the animals where the management response is to interfere with and deny animal agency as much as possible. Double-crested cormorants are one of many species whose individuals expanded their range of their own accord (rather than through human introduction) but are still labeled as 'invasives.' Their colonies, similar to cities, are often grounds of refuse, their guano so strong that it suffocates the trees on which they roost. The commercial fisheries industry along coastal Canada and northern New England has contributed greatly to their population boom and economic success. They have become so successful that they are legitimate competitors with sports anglers and commercial fishers (Wild 2012). As a consequence, state wildlife agencies enacted programs to 'manage' their populations, mostly through sterilization and killing.

Leopold (1933) was the first to mention that wildlife management is more so "people management." Kellert's work and the emergence of human dimensions of wildlife develop on this premise. However, approaches to wildlife management, as indicated throughout Leopold's own influential book *Game Management*, focus on wild animal populations as the centerpiece of management. When 'wildlife management' becomes more focused on "people management" and is a process of collaboration and transformation rather than a tool for subjugation or control, it forces stakeholders to concentrate more on taking responsibility for their actions and adjusting their behaviors before being so quick to impose on wild animal communities, especially when issues of spatial and resource conflict with wild animals arise. It also opens the possibility for interspecies collaboration.

Interspecies collaboration is a very new way of thinking about wildlife management. It is so new that, as of yet, there is no mention of it in the wildlife, environmental management, or geography journals. It is the idea that wild animals play an active, assertive role in shaping their worlds and thus can have a more participatory role in managing the human–animal interface. Gay

Bradshaw's thinking most resembles this idea though she does not use the term 'interspecies collaboration.' In her essay with Carol Buckley (founder of the Elephant Sanctuary in Tennessee, USA), they use the term 'cultural brokering' to describe "the act of bridging, linking, or mediating between groups or persons of differing cultural backgrounds for the purpose of reducing conflict or producing change" (Buckley and Bradshaw 2010: 5). They introduce this term in the context of healing PTSD in wild elephants and the psychological consequences of habitat loss and community destruction they are enduring. This concept of cultural brokering links closely with the idea of interspecies collaboration. They are both ideals that require active participation across species based on a mutual understanding in which the two species can coexist. And they both require cultural brokers or ambassadors to navigate and communicate in multiple cultural spaces at once. One of the challenges to this concept is the seemingly contradictory stance of supporting both an individual's sense of self-thriving in one culture while simultaneously insisting that the self must accommodate norms of a secondary, more dominant culture that retains control and power (Aikenhead and Jegede 1999). That is certainly a challenge that deserves deeper exploration in critical animal studies.

Sarna and Woelfle-Erskine (2013) presented at a Center for Science, Technology, Medicine, & Society talk on the role of beavers in watershed management of the Klamath Basin and the opportunities for interspecies collaboration. Many decades of persecutions and entanglements with beavers led managers to consider how they could work with beavers to improve watershed management. Beavers have been responding to management attempts on their families and communities with a mix of cooperation and resistance. By cooperating with beavers on mutual terms through processes that also allow them to affirm their lifeways and traditions, while simultaneously advancing goals that benefit whole ecological communities, interspecies collaboration begins to take shape. In essence, it is a budding opportunity to affirm the importance of agency on all sides.

One of the ways in which to acknowledge animal agency is by shifting from the animal 'question' to the animal 'condition,' whereas 'the actual life situation' of animals is the site for consideration (see White, this volume). In addition to the individual's life, the life situation for the intraspecies and interspecies community in question is just as important, for the community is pertinent to shaping animal perspective, place-based ways, and experience. Too often these qualities are completely ignored and of no consequence in conservation research. Conservation acts as a way to balance commodified resources between the processes of accumulation and loss. How can wild animals ever live in dignity and freedom in this ideological system? How can we expect to advocate for animals in such a framework? Finally, how can we build on the work that scholars and activists of color are already doing to bring people of color's perspectives and contextualized experiences into the realm of dominant decision-making and 'management?'

Extending ecological justice

The environmental justice framework is one existing alternative. By extending the principles and goals of environmental justice to include wild animals as legitimate agents deserving environmental justice, the schism between animal rights and wildlife conservation can dissolve. The environmental justice movement serves as a counter-political and ecological approach to address the life-threatening impacts of racism, classism, and colonialism in the lives of people of color and with respect to their communities. Because environmental justice is built on real-life practices of diversity and solidarity while advancing theoretical developments into interconnected patterns of oppression often rendered invisible (Taylor 2002), it is a useful framework for observing the institutional behaviors of U.S. conservation as it concerns people of color *and* wildlife. It is a living example of intersectionality where theory meets practice.

One of the early premises of environmental justice is that the lives of people of color matter and are intricately connected to the environments and ecosystems they live in. By standing up for themselves and the ecosystems they live in, people of color demonstrate their agency to shape their worlds and in turn promote life-affirming values that support all life (i.e., living diversity). Rather than motivated by aesthetic, romantic visions of the landscape, people in the environmental justice movement have been motivated by their desire to survive and thrive in a society that would, if left to its own devices, manage them into slavery or extinction. This is a concern and struggle communities of color can share with wild animal species. The stigma of associating racial minorities with nonhuman animals, particularly wild animals, and the demoralizing effects such associations have on people of color's self-concept are the biggest obstacles to bridging solidarity between people of color and wild animals. This stigma is a symptom from centuries of racist and colonialist exploitation where white-identified people continue to perpetuate imagery and narrative that relegate people of color to a ranked but inferior, oppressed social status associated with animals. Associating peoples of color with animals is not a reflection of the people or the animals, but rather an indication of just how entrenched white people's violence toward and sense of entitlement over animals was and continues to be. I realize bringing the two so closely together may be sensitive for many readers, but I take a bold step and bring these groups' histories and present struggles side by side because it is important for us to see our interconnectedness in these systems of oppression, without going so far as to trivialize or privilege one group's oppression over the other.

This is a beginning in such intersectional work, and I think this chapter only touches the surface of this vast issue. However, I hope that this will contribute to present and future collaborative work toward understanding oppression, advancing justice, generating valuable and critical knowledge concerning the human–animal interface, and solving ecological problems in the United States. In many conversations I have had with people of color deeply involved in the pursuit of environmental and economic justice, the most common value shared is that wild animals need to be 'left alone.' Although Kellert may interpret this response as uncaring or ambivalent, I think it is an opening into the shared yearning for auto-nomy, self-determination, and freedom. But rather than being rooted in a tradi-tional wildlife conservation ethic, it is placed within the context of grassroots social advancement and justice.

As the boundaries of ecological justice become more open to include wild animals, environmental justice activists may be more likely to account for wild animals who are part of their communities, as opposed to animals from species geographically and ecologically far away. Kellert (1996) stated that virtually no difference appeared between black and white people concerning domestic animals, e.g., pets and farmed animals. Black people's attitudes toward domestic animals were not significantly different, I think, because they live with these beings in community and therefore can connect on a meaningful level. There-fore, I suggest that place and community are the focal points for including animals as legitimate beings in the quest for ecological justice and the key to shifting how wildlife management in the United States is done. Being able to have regular contact with wild animals in your local community can be a reminder that we have shared vulnerabilities and interests, whether from global warming or the quest for a clean, healthy watershed. And inviting wildlife man-agement into more spaces of community-based decision-making where recogni-tion of agency, collaboration, and inclusion are prevailing virtues is a step toward decentralizing the field from overwhelming whiteness to multiethnic, interspecies solidarity.

References

Aikenhead, G and Jegede, O 1999, 'Cross-cultural science education: A cognitive expla-nation of a cultural phenomenon,' *Journal of Research in Science Teaching*, vol. 36 (3), pp. 269–287.

Allison, M 2000, 'Leisure, diversity, and social justice,' *Journal of Leisure Research*, vol. 32, pp. 2–6.

American Anti-Vivisection Society n.d., *Primates by the Numbers*. Viewed February 2, 2014. www.aavs.org/site/pp.aspx?c=bkLTKfOSLhK6E&b=7908639&printmode=1.

American Psychological Association 2012, *Ethnic and racial disparities in education: Psychology's contributions to understanding and reducing disparities*, Presidential Task Force on Educational Disparities. Viewed February 2, 2014. www.apa.org/ed/resources/racial-disparities.aspx.

Bermúdez-Cuamatzin, E, Ríos-Chelén, A, Gil, D, and Macías Garcia, C 2011, 'Experi-mental evidence for real-time song frequency shift in response to urban noise in a pas-serine bird,' *Biology Letters*, vol. 7 (1), pp. 36–38.

Bierbach, D, Girndt, A, Hamfler, S, Klein, M, Mücksch, F, Penshorn, M, Schwinn, M, Zimmer, C, Schlupp, I, Streit, B, and Plath, M 2011, 'Male fish use prior knowledge about rivals to adjust their mate choice,' *Biology Letters*, vol. 7 (1), pp. 349–351.

Bruskotter, J, Enzler, S, and Treves, A 2011, 'Rescuing wolves from politics: Wildlife as a public trust resource,' *Science*, vol. 333 (6051), p. 1828–1829.

Buckley, C and Bradshaw G 2010, 'The art of cultural brokerage: Recreating elephant-human relationship and community,' *Minding the Animal Psyche*, vol. 83, pp. 4–27.

Cooper, C 2014, 'Black students and the "School-to-Prison pipeline",' *Al Jazeera America*, January 22. Viewed February 2, 2014. http://america.aljazeera.com/watch/shows/america-tonight/america-tonight-blog/2014/1/22/black-students-andtheschooltoprisonpipeline.html.

Cooper, JF 1827, *The prairie: A Tale*, Carey, Lea & Carey, Philadelphia.

Cronon, W 1995, 'The trouble with wilderness; or getting back to the wrong nature,' in J Callicott and M Nelson (eds), *The great new wilderness debate*, W.W. Norton & Company Ltd., New York.

Dececco, J and Cooper, R 1996, 'Shorebird migration at a Mississippi River wastewater treatment plant,' *Proceedings of the Southeastern Association of Fish and Wildlife Agencies*, vol. 50, pp. 221–227.

Dyer, T 1992, *Theodore Roosevelt and the idea of race*, LSU Press, Baton Rouge, LA.

Floyd, M 1999, 'Race, ethnicity, and use of the national park system,' *Social Science Research Review*, vol. 1, National Park Service, pp. 1–23.

Forehand, B 1996, 'Striking Resemblance: Kentucky, Tennessee, Black Codes and Readjustment, 1865–1866,' Masters Theses and Specialist Projects, Paper 868.

Fox, S 1981, *John Muir and His Legacy: The American conservation movement*, Little, Brown and Company, Boston.

Gilbert, F and Dobbs, D 2001, *The philosophy and practice of wildlife management*, 3rd Edition, Krieger Publishing Company, Malabar, FL.

Green, A 1996, *Animal underworld: Inside America's black market for rare and exotic species*, Public Affairs, New York.

Guha, R 1989, 'Radical American environmentalism and wilderness preservation: A third world critique,' *Environmental Ethics*, vol. 11 (1), pp. 71–83.

Harry, J, Gale, R, and Hendee, J 1969, 'Conservation: An upper-middle class social movement,' *Journal of Leisure Research*, vol. 1 (2), pp. 255–261.

Hays, S 1989, *Beauty, health, and permanence: Environmental politics in the United States 1955–1985*, Cambridge University Press, New York.

Hernandez, F and Guthery, F 2012, *Beef, brush and bobwhites: Quail management in cattle country*, Texas A&M University Press, College Station, TX.

Holt, M, Noren, D, Veirs, V, Emmons, C, and Veirs, S 2009, 'Speaking up: Killer whales (*Orcinus orca*) increase their call amplitude in response to vessel noise,' *Journal of the Acoustical Society of America*, vol. 125 (1), pp. EL27–EL32.

Hornday, W 1913, *Our vanishing wild life: Its extermination and preservation*, Charles Scribner's Sons, New York.

Izzo, J 2010, 'PC pets for a price: Combating online and traditional wildlife crime through international harmonization and authoritative policies,' *William and Mary Environmental Law and Policy Review*, vol. 34 (3), pp. 965–998.

Jacobson, C, Organ J, Decker, D, Batcheller, G, and Carpenter, L 2010, 'A conservation institution for the 21st century: implications for state wildlife agencies,' *Journal of Wildlife Management*, vol. 74 (2), pp. 203–209.

Jordan, C and Snow, D 1992, 'Diversification, minorities, and the mainstream environmental movement,' in D Snow (ed.), *Voices from the Environmental Movement: Perspectives for a new era*, Island Press, Washington, DC.

Kellert, S 1996, *The value of life: Biological diversity and human society*, Island Press, Washington, DC.

Laland, K and Galef, B 2009, *The question of animal culture*, Harvard University Press, Cambridge, MA.

Laland, K and Janik, V 2006, 'The animal cultures debate,' *Trends in Ecology & Evolution*, vol. 21 (10), pp. 542–547.

Leonard, J 2007, 'Fishing and hunting recruitment and retention in the U.S. from 1990 to 2005,' *Addendum to the 2001 National Survey of Fishing, Hunting, and Wildlife-Associated Recreation*, U.S. Fish and Wildlife Service, Arlington, VA.

Leopold, A 1933, *Game management*, Charles Scribner's Sons, New York.

McCool, S 2008, 'Challenges and opportunities at the interface of wildlife viewing marketing and management in the 21st century,' in M Manfredo, J Vaske, P Brown, D Decker, and E Duke (eds), *Wildlife and society: The science of human dimensions*, Island Press, Washington, DC.

National Hunting Club Listing 2014, viewed February 2, 2014. www.huntclublisting.com/index.php?page=modules/search/allresults.

National Survey of Fishing, Hunting, and Wildlife-Associated Recreation 2011, U.S. Department of the Interior, U.S. Fish and Wildlife Service, and U.S. Department of Commerce, U.S. Census Bureau.

Nicholson, K, Krausman, P, and Merkle, J 2008, 'Hypatia and the Leopold standard: Women in the wildlife profession 1937–2006,' *Wildlife Biology in Practice*, vol. 4 (2), pp. 57–72.

Organ, J and Batcheller, G 2008, 'Reviving the public trust doctrine as a foundation for management in North America,' in M Manfredo, J Vaske, P Brown, D Decker, and E Duke (eds), *Wildlife and society: The science of human dimensions*, Island Press, Washington, DC.

Parks, S, Johnson, M, Nowacek, D, and Tyack, P 2011, 'Individual right whales call louder in increased environmental noise,' *Biology Letters*, vol. 7 (1), pp. 33–35.

Patricelli, G and Blickley, J 2006, 'Avian communication in urban noise: Causes and consequences of vocal adjustment,' *Auk*, vol. 123 (3), pp. 639–649.

Roosevelt, T 1893, *The wilderness hunter*, G.P. Putnam's Sons, New York.

Roosevelt, T 1927, *The works of Theodore Roosevelt, volume II*, H Hagedorn (ed.), Charles Scribner's Sons, New York.

Ruland, P 2012, 'Parks, wildlife, and racism,' *The Austin Chronicle*, August 10. Viewed August 12, 2012. www.austinchronicle.com/news/2012-08-10/parks-wildlife-and-racism/all/.

Sarna, D and Woelfle-Erskine, C 2013, 'Dam 'em all: Beaver believers, beaver deceivers, and other watershed entanglements,' talk presented at the *Center for Science, Technology, Medicine, & Society Symposium*, May 10, 2013. Viewed August 20, 2013. www.youtube.com/watch?v=9IcMy8W_68M.

Schuett, M, Scott, D, and O'Leary, J 2008, 'Social and demographic trends affecting fish and wildlife management,' in M Manfredo, J Vaske, P Brown, D Decker, and E Duke (eds), *Wildlife and human society: the science of human dimensions*, pp. 18–30, Island Press, Washington, DC.

Slabbekoorn, H and den Boer-Visser, A 2006, 'Cities change the songs of birds,' *Current Biology*, vol. 16 (23), p. 2326–2331.

Slabbekoorn, H and Ripmeester, E 2008, 'Birdsong and anthropogenic noise: Implications and applications for conservation,' *Molecular Ecology*, vol. 17 (1), pp. 72–83.

Smith, K, Behrens, M, Schloegel, L, Marano, N, Burgiel, S and Daszak, P 2009, 'Reducing the risks of the wildlife trade,' *Science*, vol. 324 (5927), pp. 594–595.

Spence, MD 2000, *Dispossessing the wilderness: Indian removal and the making of the national parks*, Oxford University Press, New York.

Subcommittee on Fisheries, Wildlife, and Oceans of the Committee on Natural Resources

U.S. House of Representatives 2008, *Impacts of U.S. Consumer Demand on the Illegal and Unsustainable Trade of Wildlife Products.* Viewed September 22, 2014. www.gpo. gov/fdsys/pkg/CHRG-110hhrg44484/html/CHRG-110hhrg44484.htm.

Taylor, D 1997, 'American environmentalism: The role of race, class, and gender in shaping activism 1820–1995,' *Race, Gender, & Class*, vol. 5 (1), pp. 16–62.

Taylor, D 2002, *Race, Class, Gender, and American Environmentalism*, General Technical Report PNW-GTR-534, U.S. Department of Agriculture, Forest Services, Pacific Northwest Research Station, Portland, OR.

Texas Disposal Systems 2011, viewed October 23, 2013. www.texasdisposal.com/index. php/exotic-exotic-game-ranch.

Thoreau, HD 1854, *Walden; or, life in the woods*, Ticknor and Fields, Boston.

Tober, J 1981, *Who owns the wildlife? The political economy of conservation in nineteenth-century America*, Greenwood Press, Westport, CT.

Trefethen, J 1966, 'Wildlife regulation and restoration,' in H Clepper (ed.), *Origins of American conservation*, The Ronald Press Company, New York.

Vardanis, Y, Klaassen R, Standberg, R, and Alerstam, T 2011, 'Individuality in bird migration: routes and timing,' *Biology Letters*, vol. 7 (1), pp. 502–505.

Warren, L 1997, *The hunter's game: Poachers and conservationists in twentieth-century America,* Yale University Press, New Haven, CT.

Whiten, A and van Schaik, C 2007, 'The evolution of animal "cultures" and social intelligence,' *Philosophical Transactions of The Royal Society B*, vol. 362 (1480), pp. 603–620.

Wild, D 2012, *The double-crested cormorant: Symbols of ecological conflict*, University of Michigan Press, Ann Arbor, MI.

Wildlife Systems, Inc. 2008, viewed February 2, 2014. www.wildlifesystems.com/about_ us.asp.

Wolch, J 1998, 'Zoöpolis,' in J Wolch and J Emel (eds), *Animal geographies: Place, politics and identity in the nature-culture borderlands*, Verso, New York.

8 Pit bulls, slavery, and whiteness in the mid- to late-nineteenth-century U.S.

Geographical trajectories; primary sources

Heidi J. Nast

Introduction

According to the recollections of distinguished dogmen compiled and published by the New York-based aficionado of blood sports, Richard K. Fox (1888), U.S. dog fighting was introduced along the eastern seaboard as early as 1816, nearly two decades before the British Parliament banned dog fighting in 1835. It was primarily white working-class men from the coal mining and manufacturing centers of Great Britain who carried the 'pit bull' to North American shores. Created by crossbreeding terriers with bulldogs, the bull terrier hybrids evinced the best of both breeding worlds: the tenacity, speed, and prey drive of terriers, and the strength and jaw hold of the old bulldog. Most fighting dogs of the century weighed between twenty-five and thirty-five pounds, smaller than those of later centuries made by outbreeding with larger mastiff and alaunt kinds (Armitage 1935).

Formal fights in both Britain and the U.S. were staged in show rings (pits) and involved a slew of others, including owners, stakeholders, referees, medical help, trainers, and timekeepers. Dog fighting's similarity to the blood sport of bare-knuckle boxing was not coincidental, but tied closely to the lives and struggles of the working class out of which both sports emerged (Nast 2013). Decades later, a second domain of canine spectacle would emerge, this time in bourgeois circles throughout major British and U.S. cities: the show ring. While the first ring was based on use value, the second was founded on exchange value, a differentiation that shaped the respective 'how' and 'why' of breeding.

Fighting dogs, for instance, were not expected to look the same but, rather, were bred and trained to fight, a form of struggle with which the working class expressly identified. The best dog fighters, it was believed, were those bred from fight champions themselves. Outbreeding was hence the norm. By contrast, the bourgeois show ring was a place where race purity was celebrated, produced through the practice and language of standardization and conformation.[1] The best dogs were those inbred for generations to make what were called pure breeds or purebreds. Judges awarded the highest points to those exhibiting the highest degree of racial sameness, with demand for certain aesthetics determining market value (Nast, forthcoming).

Primary and secondary sources suggest that white settlers in the U.S. introduced fighting dogs into four distinct, if overlapping, colonial domains: the coal mines of Pennsylvania and Ohio, the factories of the northeast, southern slave plantations, and the expanding frontier, in each case the dogs being called upon to attend to particular needs and desires of white settler life. They might be relied upon, for instance, to kill rats or other vermin, help in the hunting of boar, protect homesteads, and/or catch slaves. The moniker 'pit' came from the coal pits of Britain where pitted dog fighting first emerged and from where dog fighting migrated to pits (or fighting rings) built in the backs or basements of rural and urban pubs, inside public arenas (in Britain, only), and set up informally in open areas such as fields or alleys. The diversity of the dogs' British provenance and regional purposes to which the dogs were put in the U.S. are evident in the many informal names given the 'breed': pit terrier, pit bull terrier, half-and-halfs, Staffordshire fighting dog (from England), Old Family dog (an Irish name), Yankee terriers (used by northerners), and rebel terriers (used by southerners), to name a few.[2] Formal 'pit bull' breed names began assuming cultural importance during and after the 1860s when various breeders worked to standardize and proprietarily coin a 'fighting' look for the show ring (Nast, forthcoming).

Despite show dog fanciers' depiction of fighting canines as degenerates and mongrels, dogmen involved in the blood sport considered their dogs to be an elite breed apart. The class-striated interests separating the show ring and fighting pit meant that the pit bull occupied an ambiguous position within the emerging modern dog world, one dominated by whites. Such ambiguity is still evident in the many ongoing and acrimonious debates about the criteria of 'true' breeds and the related arguments about whether or not the term 'pit bull' refers to a single separate breed, many breeds, or no breed at all. I do not engage in these debates here, but rather follow in the footsteps of the dogmen themselves as they organized and claimed the racialized landscapes of early dog-fighting life.

This paper focuses on one of the four trajectories delineated above, namely, southern plantation life, an arena heretofore undocumented in animal studies work. In so doing, I present new primary material that points to how fighting dogs and dog fighting were instrumental in asserting the racial truths and consequences of white supremacy. The sources include: an 1868 Civil War court case; interviews from 1936 and 1938 with former slaves' descendants conducted as part of the Federal Writers' Projects (FWP); a monograph published in 1938 about antebellum New Orleans life written as part of the Work Projects Administration (WPA); several late-nineteenth-century lithographs printed by Currier & Ives that racistly portray black men attempting to dog fight; news articles about pit bulls being pitted against black men in 1901 and 1919; short nineteenth- and twentieth-century articles in white and African-American newspapers; and World War I war posters. The emphasis on primary materials allows me to complicate ideas circulated today in mainstream 'white' media that imply that dog fighting is something carried out mostly by men of color. The hysteria over Michael Vick is perhaps the most salient example of this (Massey 2012).

Before beginning a discussion of these sources, however, I briefly overview the importance of fighting dogs to rural frontier life, more generally, where settlers westward-bound shared some of the welfare and security concerns as those who were settling (or had already settled) the south, including plantation owners. For both, the dog proved especially important in ridding farmlands of a major pest, the wild boar. While it is unclear to what extent frontier folk on small farms deployed the dogs for fighting purposes, primary evidence exists that miners and prospectors from Colorado to California used the dogs for blood sport, as did southern slave owners who had effectively settled a related frontier. The dogs' varied uses were hence entangled in the larger project of colonial expansion.

Frontier farming and wild boar hunting in the southern U.S.

Southern farmers and frontier folk used pit bulls for many similar reasons. White settlers that established ranches, farms, and homesteads in either context used the dogs to retrieve stray hogs and cattle and/or for protection. The varieties of dogs used were generally larger and leggier than those first brought to the U.S. from Ireland and England.[3]

One of the most regular threats to southern farming life for which the large size of the dogs proved critical was the wild boar. Living primarily in the southern and western U.S., boars regularly wrought havoc on farm fields owing to their method of feeding, called rooting. Tusks lowered into the ground, they run lines back and forth, eating just about everything in their path: grass, nuts, acorns, berries, earthworms, roots, tubers, and insects as well as rodents, reptiles, amphibians, carrion, and small mammals. They accordingly can destroy not only crops, but the soil's structure and the food sources used to sustain micro-ecologies useful to farm life.[4] Due to their large size and aggressive nature, boars were also dangerous and regularly hunted. At the same time, wild boar meat was considered delectable such that wild boar hunting became a prized activity in and of itself.

The strong prey drive, tenacity, alacrity, strength, and powerful jawline and grip of pit bulls made the dogs indispensable to wild boar hunts. The dogs became renowned for the boldness with which they could chase and take hold of ('catch') an animal many times their size. Their superior capabilities made them distinguished catch or 'catch and hold' dogs, of which there were (and are) several, all of them derived from the bulldog.[5] Over time, hog hunting with pit bulls ('hog and dog' events) became part of a well-established (male) tradition of the rural, white south, with wealthier interests later in the century introducing European boar varieties to give the hunt a more elite feel.

The ability of pit bulls to pursue and catch large animals attracted slaveholders, who began using the dogs to track runaway slaves or paid bounty hunters to do the same (below). The regionalized ties between southern boar hunting, pit bull fighting, and slave catching remain as a discursive sediment in the language and imagery of contemporary southern 'dog and hog' events, which can be seen today on YouTube videos and are discussed on a variety of blogs, websites, and vanity-published books (e.g., Adele 2012; Kelley 2009). In most cases, details of a hunt are relayed by young white male hunters who acclaim their exploits, centering on the ferocity of their dogs, the doggedness and massive size of the boar (today's varieties can weigh over 900 pounds), and the skillful collaboration between dog and man in carrying out a kill. Implicit in many of these accounts is a sense that such hunting is a southern white man's preserve. The background image of a Georgia man's website for his 'red nose American pit bull terrier' business, for instance, features a Confederate flag in front of which is seated a paired "dog [pit bull] and hog [boar]" image. One of the site's links takes you to a "pit bull and rebel wear store" that sells "stuff for pit bull lovers, hunters, and proud southerners."[6] I mention the boar's importance to both frontier life and the rural south as a segue to discussing the additional and special importance that pit bulls assumed in the context of southern slave-holding.

Slave-holding and the pit bull

In 1868, a Congressional military tribunal charged Confederate Major Henry Wirz with "murder, in violation of the laws of war," for which he was convicted and later executed. The murder charge stemmed from his tenure as commandant of a Confederate prison known as Camp Sumter, located in Andersonville, Macon County, Georgia (U.S. House of Representatives 1868; National Park Service 2014). The camp, built in 1864 and enclosed by a fifteen-foot high stockade, was supposed to house about 10,000 prisoners but at one point held more than 33,000. By the time it was liberated in May 1985, over 45,000 persons had passed through its gates, roughly 13,000 dying of scurvy, diarrhea, and dysentery. Because food was scarce, many Union prisoners starved to death. While 351 prisoners attempted escape, many were recaptured, the U.S. Army recording only thirty-two persons who successfully reached Union lines, though others may have returned home without re-registering.

Survivors called to testify about Major Wirz's actions recalled how he deployed two distinct kinds of dogs to track runaway prisoners. One survivor remembers that:

> [Harris, a slave bounty hunter] had a pack of eight hounds, besides a dog which they called a 'catch dog.' That dog always went with the pack.... Turner [another slave bounty hunter] tended about fifteen dogs.... I saw one young man who had made his escape.... The young man ... was very weak when he made his escape ... and then he was torn by the dogs and was much weakened by the loss of blood ... his legs were torn so that he could not walk, and his shoulders and neck were torn and his clothing was nearly all torn from him.... [T]he dogs overtook him and he climbed a tree, and he said that this old gentlemen, Harris, and Captain Wirz shook the tree so that he fell down and allowed the dogs to tear him.... I understand that he died that night.
>
> (U.S. House of Representatives 1868: 320)

Because pit bulls had, by this time, not been standardized, the witnesses struggle to describe the breed characteristics of the catch dogs. Hence, one witness remarks that, "I saw the dogs while I was there.... They had also one large dog, which, I think, they called 'catch-dog'; I think he was a bull-dog, or a bull-terrier of some kind" (U.S. House of Representatives 1868: 60; similar comment on p. 97). Many persons of the time identified fighting dogs as either bull dogs or bull terriers (which was also the case in Britain). Another man describes the catch dogs as brutal when he states that "there were a couple of them called catch-dogs; the others were hounds.... I believe there are track-hounds and catch-dogs.... One is very vicious by nature.... The catch-dog is vicious" (U.S. House of Representatives 1868: 131). A third witness confirms the catch dog's ferocity, when he explains that:

> there were two kinds of dogs, the hound and the catch dog, as he was called; I guess he was a bull terrier or something of that kind.... He was more ferocious than the hounds; I saw one man who had been bitten by the [catch] dogs ... the calf of his leg was torn pretty nearly off.
>
> (U.S. House of Representatives 1868: 142)

Analogous testimonies abound.[7]

The statement of a prisoner who formerly worked as a slave bounty hunter in Virginia teaches us not only about the practical utility of the scent-catch distinction, but of its human-oriented use in *slavery*. He begins by relaying that, "I escaped from the stockade some time in August.... I took to the tree" (U.S. House of Representatives 1868: 401). He goes on to note the importance of trees to runaway slaves; they provided relief from the dogs and presumably also a place to which a slave could retreat for safety, as needed. The bounty hunter hence says:

> [Before] I had [had] a good deal of fun catching niggers.... For that reason I went up a tree. I knew what they [the runaways] would do. When I came down the tree they [the hounds] did not bite me, for the catch-dog was tied. If he [the catch dog] took hold they [the hounds] would all take hold.... I caught negroes in Hardie county, Virginia. I used to have real good sport there at it.
>
> (U.S. House of Representatives 1868: 401)[8]

Tree-climbing was not always successful, however. Not only could a tree be shaken to make a runaway fall out of its branches, as noted above, but the high prey drive of bull terriers meant that the dogs could *climb* the trees in pursuit – or at least 10′ to 15′ worth. Once ensconced in a tree's crotch, the dogs could restart their ascent. (Some pit bulls today climb trees when pursuing tree rats or cats.)

The importance of trees to runaway slaves is corroborated in firsthand accounts of former slaves collected between 1936 and 1938 by writers working for the Federal Writers' Project (FWP). Will Glass, a paver working in Arkansas, tells his FWP interviewer about one of his uncles who had an especially "tough master":

> He was so tough that Uncle Anderson had to run away. They'd whip him and do around, and he would run away. Then they would get the dogs after him and they would run him until he would climb a tree to get away.... They would come and surround the tree and make him come down.... Sometimes *they would make the dogs bite him* [hounds would follow the lead of the catch dog] and he couldn't do nothing about it. One time, he bit a dog's foot off.... They whipped him again. They would take him home at night and put what they called the ball and chain on him.
>
> (FWP 1936–1938a: 38, italics added)[9]

A fight notice posted on May 16, 1856 in the twice-daily New Orleans' newspaper, *The Daily True Delta*, tells us that such dogs were likewise used in pitted fighting for entertainment purposes. The notice refers to two New Orleans dogs:[10]

> Match for $500. – Great Sport for Dog Fanciers. – The longed talked of match between the celebrated fighting dogs, Spot and Jack, belonging to two gentlemen of this city, will take place at the American Cockpit, on Gravier-street, opposite the Gaiety Theatre, on SUNDAY, the 18th inst., at 12 o'clock, precisely. Admission – One Dollar.
>
> (*The Daily True Delta*, May 16, 1856)

The Daily True Delta carried mostly political and business news, an important part of which were its listings of slave sales and slave bounty amounts, which ranged that day from $20–$25.

By this time New Orleans had become the fourth largest U.S. city, second only to New York as a port. This, despite the fact that the new Erie Canal and railway lines, which ran from the east coast to the Midwest, had begun diverting thousands of tons of goods out of the Ohio Valley and towards New York. Even in 1846, when flour and wheat receipts in Buffalo exceeded those in New Orleans, the rise in cotton and slave-trading receipts helped the city thrive. Indeed, by 1850 New Orleans had become the "greatest slave market in the country" as well as "the cultural center of the South" (WPA 1938: 20):

> Opera flourished, theaters attracted European stars, artists abounded, and *bon vivants* thrived in a city ... already. .. famous for its fast and loose manner of living. Gambling, horse-racing, dueling, steamboat racing, and cock- and dog-fighting, in addition to the magnificence of balls, receptions, and Mardi Gras, made New Orleans ... a gay metropolis.
>
> (WPA 1938: 21)

The fighting pits' location is important in that Gravier Street had by now become one of two important slave-trading centers in New Orleans. Its formal presence and placement accordingly repeated and reinforced metonymically a key relation between catch dog and slave.

Firms famously kept their African cargo in pens surrounded by high walls, not unlike those that surrounded Camp Sumter,"[t]he walls ... were so high—fifteen or twenty feet—that one New Orleans slave dealer thought they could keep out the wind" (Johnson 1999: 3). Both fighting pit and slave pen served settler purposes. In particular, the spectacle of gladiatorial canine combat celebrated by the one spoke of the brutal force of supremacist expansion that allowed for the containment and exploitation of the other.

Following the Civil War and emancipation, African-American men began tenuously to enter the formerly whites-only dog-fighting fray. To wit, in northern dog-fighting circuits, we see rare allusions to 'negro' fight participants, almost all of them cited as 'trainers.'[11] This may have had to do with the fact that training fighting dogs involves considerable physical rigor for both trainer and dog, some early trainers recommending that the dogs be walked for miles, daily. The arduousness involved for both dog and human suggests that training in the antebellum south would have been the preserve of slaves, a poignant irony given their slave catch functions.

Occasional mention is also made of African-American dogmen operating in 'Negro-only' settings, to which the white community paid scant attention. In 1898, for instance, the conservative New York newspaper, *The Sun*, included a few lines about Nevada Senator Stuart who owned a "bull pup" [pit bull] "of which he is very fond" (*The Sun*, March 27, 1898: 3). One evening the dog disappeared, which the senator reported to the police:

> The next day a policeman saw a colored man with the suspected animal.... It then developed that the bull pup had been stolen ... and had engaged in a dog fight.... The dog had whipped his antagonist and won $25 for the thief.
>
> (*The Sun*, March 27, 1898: 3)

Given the poverty of ex-slaves, fight purses tended to be small, the fights informal, and the spaces marginal. Fights commonly took place in alleys, streets, or in out-of-the-way areas. The 'Police Court' section of the March 19, 1901 *Alexandria Gazette and Virginia Advertiser* (p. 1), for instance, notes that "Frank Baltimore, colored, charged with fighting dogs in the street, was dismissed. Lina Belden, colored charged with the same offense, was dismissed." In 1904, the Maysville, Kentucky's, *The Evening Bulletin* (May 17, 1904: 1) had a short entry about an unusually large gathering of "100 Negroes" for a pit fight between two dogs, Jim and Emma, "just outside the city limits" of Louisville, Kentucky. The dogs refused to fight, which led to a huge row and two fatalities. The paucity of funds available for fight purses is likewise evident in a complaint published in 1910 in the African-American newspaper *The Topeka Plaindealer* (September 9, 1910: 5) under the heading, "Using the Church for a Bill Board!" According to the anonymous editorial:

The Negro churches of this city are ... called upon to advertise ... and to announce everything from a dog fight to a love feast.... Negro business institutions ought to be required to advertise their specials through the newspaper.... Colored preachers should put an end to this announcing business.... [T]he pastor who keeps it up is either a coward or a man hunting cheap favors from stingy people.

(*The Topeka Plaindealer*, September 9, 1910: 5)

Between 1904–1907, one African-American newspaper, the *St. Louis Palladium*, published several short news items and commentaries about "Negro" dog fighting, none of these appearing as anything other than trifling (if regular) events.[12]

Post-Civil War white supremacy and the pit bull

Between the mid-1870s and early 1890s, two artists working for the renowned Massachusetts-based print-making firm of Currier & Ives produced *Dark Town*, a racist lithograph series depicting the civilizational ineptitude of recently freed slaves. Along with other popular media of the time – such as *Harper's Weekly* (which produced the denigrating print series, *Blackville*), *Life*, *Puck*, *Judge*, *New York World*, *New York Journal*, *The Outlook*, and *The Independent* – Currier & Ives worked actively to communicate the superiority of the white race (and the white man, in particular). This was a time of considerable white hysteria and the ushering in of the post-Reconstruction order. It involved working towards the political and economic disenfranchisement of newly freed slaves, elaborating Jim Crow laws and practices, creating the prison and sharecropping system, establishing the Ku Klux Klan (KKK), the carrying out of hundreds of yearly lynchings, and projecting the violence of white actions onto the body of the black man. Within this context, the *Dark Town* images were a way of calming white fears by giving them the imaginary means by which to consider black folks, especially men, as fundamentally incompetent, foolish, and simple. Two of the *Dark Town* images use a dog-fighting event to stage these facts.

The first image shows a short, squat, cigar-smoking African-American man with enormous lips wearing a vested suit with a red neck cravat (see Figure 8.1). Both his hands hold onto a short metal chain-leash at the end of which a white pit bull dog strains. Immediately behind him is an untidy group of young black men, all in various degrees of suited attire and wearing some kind of top hat or other. One of the men smokes a cigar (a sign of wealth and white prerogative) and wears white gloves. The title (*A Sure Thing*) and subtitle (*It's Picking up Money Backing Dis Yere Pup*), along with the racial ungainliness of the participants and the undisciplined attitude of the dog is intended to communicate the outlandishness and ignorance of their pursuit, one that only white men can properly do.

The title and subtitle of the second image – *All broke up* and *He'd a Won De Money, if It Hadn't Been for De Odder Dog* – shows the ragtag group after the fight (see Figure 8.2). It is now two fewer in number and accompanied by a haggard black woman, her hair covered in a polka-dot kerchief, the tied ends of

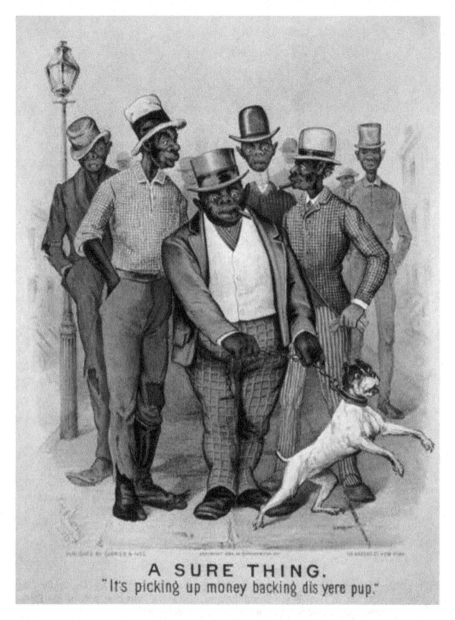

Figure 8.1 A Sure Thing: It's Picking up Money Backing Dis Yere Pup (Currier & Ives, NY, chromolithograph, *c*.1884, issued as part of its 'Darktown comics' series, Library of Congress archives)

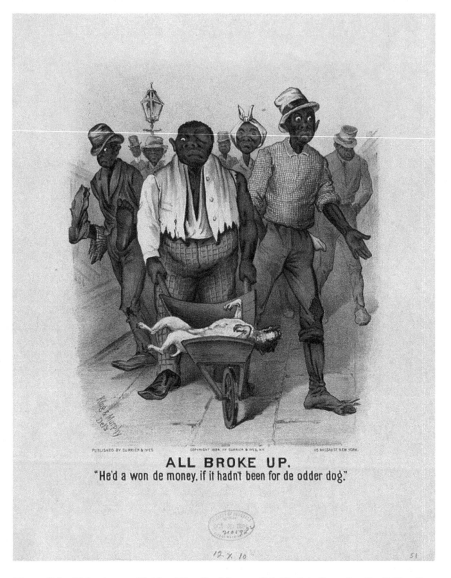

ALL BROKE UP.
"He'd a won de money, if it hadn't been for de odder dog."

Figure 8.2 All broke up: He'd a Won De Money, if It Hadn't Been for De Odder Dog (Currier & Ives, NY, chromolithograph, *c*.1884, issued as part of its 'Darktown comics' series, Library of Congress archives).

which point upward. The main figure has lost his top hat along with one of his ears. His face is swollen and he has a black eye, presumably the result of getting mixed up in the fight itself or in a post-fight fray. His suit coat and vest are in tatters, and his tie is gone. Instead of walking the dog, he pushes it in a wheelbarrow where it lies belly-up, dead, the obvious loser. The rest of the group follows him, undone, bemused, downcast. The formerly one-gloved man is injured. Missing a shoe, he is carried by the others.

The message is clear: dog fighting is a decidedly white masculine endeavor, at which African-American men are bound to fail. The disappearance of all the signifiers of whiteness (cigars, shoes, cravat), the exaggerated physical features of the participants, and the death of the dog make the attempt appear farcical, an exercise in buffoonish mimicry. The lithographs thus help render black men impotent at a time when emancipation had just set them free (see Nast 2000).

In 1889, Currier & Ives published a similarly hateful print, *De white dog's got him!* It centers on a rough-hewn two-sided dog-fighting pit peopled by a waggish crowd of African Americans (see Figure 8.3). A tiny basin sits in the pit (used to wash fighting dogs), and the room is lit by candles on the floor and on a board suspended from the ceiling. On the left side of the pit stand nine black men and boys with apelike faces who holler, arms akimbo, as their pure white dog gets the better of the opponent's pure black dog. Two of the men wear top hats, while another wears a fez cap, the kind that monkeys wore when collecting tips for itinerant organ grinders. On the opponent's side of the pit are eight men and boys who appear comically perplexed as they regard their losses. One of them wears a baseball cap (first invented and worn by white baseball teams in New York City in the 1860s) while two others wear formal and expensive suits with coat tails. The constellation of signifiers that in white hands communicates civilizational qua racial prowess is remade to signify racial ineptitude, proof of the impossibility of black manhood or humanity.

DE WHITE DOG'S GOT HIM!

Figure 8.3 De white dog's got him! (Currier & Ives, NY, chromolithograph, *c.*1889, issued as part of its 'Darktown comics' series, Library of Congress archives).

Taken together, the three dog-fighting-centered images speak of white anxieties over emancipation, Reconstruction, and African-American enfranchisement. The work that Currier & Ives sets out to do is limited but important: to portray a common humanity as impossible. Further, in positioning black bodies as fundamentally incapable as they don the material mantles of, and carry out activities associated with, whiteness, the white viewer is inoculated against the shock of seeing African Americans engaging freely with a world that was once proprietarily their own. In other words, despite the law, African Americans are imposters.

Rendering black men as fundamentally inferior to white men (and their dogs) was correspondingly made plain in direct physical ways that reasserted the truths of the old order, one in which white men used dogs as instruments of racist power. In 1901, for instance, a white man in Newport News, Virginia forced "a 15-year-old Negro to fight a bull dog by threatening him with death. The dog was turned loose and the boy's hands were badly chewed up."[13]

A similar incident happened in 1919, a time when African-American hopes for a just future following their service in World War I were dashed by violent reassertions of white supremacy, including the efflorescence and expansion of the KKK into the Midwestern and northwestern U.S. The event took place on a hot day in August in Vicksburg, Mississippi:

> where a few days ago thousands of white men and women dug a hole in the ground and buried a colored man, all but his head and then, after tantalizing a vicious bulldog to a frenzied state, put him in the iron cage that covered the head of the doomed man and then danced with delight while the maddened brute tore the man's tongue and eyes out and likewise scalped him.
>
> (*Cayton's Weekly*, August 1, 1919: 2)

The atrocity speaks to how supremacist violence continued to be celebrated as virtuous, white, and right; a means of maintaining the civilized-uncivilized distinction. We see this further in how the U.S. depicted itself on World War I postcards as a pure white, fighting pit bull dog, ears clipped close to its head (unclipped ears were a liability in fights). Prior to U.S. entry into the war, the dog is festooned by a U.S. flag tied around its neck. One postcard caption reads, *Watchful-Waiting*, a phrase connected to President Wilson's early neutrality policy adopted in 1914 (see Figure 8.4). After the U.S. joined the war in 1917, a new postcard was issued depicting the dog in a U.S. Navy uniform, an Army division that President Theodore Roosevelt had expanded as part of his imperial ambitions in the Caribbean and Central America, following U.S. Navy-led incursions into Cuba and the Pacific (see Figure 8.5; Reckner 2001; Morris 1979).

What all of these images point to is the flexible nature of supremacist logic: anything that whites do – including dog fighting – is civilized and exceptional; anything nonwhites do is 'not.' African-American forays into pit fighting after the Civil War threatened this logic, as did all other efforts to enfranchise African

Watchful-Waiting

THE STAR SPANGLED BANNER

Three Cheers for the Red, White and Blue

THE AMERICAN WATCH-DOG

We're not looking for trouble
But we're ready for it.

Figure 8.4 Watchful-Waiting (World War I postcard printed *c.*1914 prior to U.S. entry in the war, Henry Heininger Co. Publishing, a postcard publisher that shut down sometime in the 1920s).

Figure 8.5 The American Watch-Dog (World War I postcard produced *c.*1917 after the U.S. entered the war, Henry Heininger Co. Publishing, a postcard publisher that shut down sometime in the 1920s).

Americans. In the absence of slavery, discursive constructions of inferiority assumed great importance, attempting to rescue a supremacist order invested in refusals (to deal with the past), abjection, and poverty.

Reagan and the 1980s

I do not have the time or space, here, to present and discuss the data that traces men of color's involvement in twentieth-century dog fighting in the U.S. In general African-American investment in the blood sport remained negligible until the global recession of the 1980s, when President Reagan cut public funding for housing and public welfare. Unemployment rates amongst African-American men skyrocketed as they became the first to be laid off, and survivalist drug, prostitution, and arms economies took hold. Pit bull dogs were used to protect crack stashes and drug dealers, and served as fighters in pit events around which various actors involved in survivalist enterprises gathered and gambled.

Like the pit-itself, the ghetto had become a space characterized by intense degrees of struggle largely determined by those beyond its confines.

The largely ghetto-bound efflorescence of African-American dog fighting in the 1980s begs many questions, not least, what it meant to those participating. Unlike in supremacist contexts, the dogs were not being deployed as subordinating instruments of white power. Rather the dogs served as important income generators in their own right and as security devices within a world of precarity, similar to how they operated initially amongst the white working-class poor (Nast 2013). That said, did those invested in dog fighting identify with the dogs? Disidentify? Were fights a way of pitting one denigrated figure against the other in a public staging of social Darwinist struggle with which spectators and participants themselves identified? Were they a way of showing the possibilities of survival despite impossible odds, but that survival only came at the other's expense? This line of questioning is crucial if we are to understand how any dog-fighting practices come to pass. The questions also bring me to the case of Michael Vick for whom the ghetto and dog fighting were very real.

Lost opportunities

On December 28, 2010, Tucker Carlson, while guest hosting Sean Hannity's show on Fox News Channel, called for the execution of African-American Falcons player Michael Vick for his participation in dog fighting. A white conservative and self-professed Christian, Carlson made his remarks as part of a critique of Obama who had recently applauded Philadelphia Eagles owner Jeff Lurie for signing Vick onto his team following the latter's eighteen-month incarceration at Leavenworth federal penitentiary (ESPN 2010).

When news of Vick's Bad Newz Kennel surfaced in 2007, much of the attention was focused on the more than fifty fighting dogs found there. The condition of the dogs was used as proof that Vick was monstrous, an unsympathetic figure whose fundamentally depraved nature led him to a love of blood sports (Broad 2013; Massey 2012). In so doing, the media drew on centuries of racist depictions of black men as uncivilized, in this case, lower than the dogs that had once been used against them. PETA posited that Vick was biologically inferior and called on the NFL to submit Vick to psychological testing and an MRI brain scan to "help determine if Vick can ever truly understand that dog fighting is a sick, cruel business" (Broad 2013: 787). Most whites called for no mercy, while many African Americans called for historical context and understanding, even if most persons polled claimed that some kind of jail time was due. Vick was additionally disciplined by the Humane Society of the United States (HSUS), becoming their spokesperson for tougher antidog-fighting laws (e.g., HSUS 2011).

Throughout the investigations and trial, no interest was evinced in why the blood sport emerged in the first place or its historically and geographically varied nature. Following Michael Vick's arrest, the media instead attended to the innocence and impoverished state of the dogs as proof of the moral decrepitude of Vick and others in his dog-fighting circles. The media also showed great interest in documenting the white heroism of state and nonprofit actors involved in various exercises of rescue. A slew of stories hence ensued about adoptions and rehabilitation, whiteness becoming bathed in a light of moral goodness (e.g., Schulte 2008; Strouse 2009; Gorant 2010; Seven Survivors 2012; Sieczkowski 2014). Analytically speaking, the circumstances of the blood sport and Michael Vick's life were disappeared, the varied racial and imperial logics of the blood sport buried in the media rubble. Lost were connections between past and present, between Vick's life and U.S. pit bull history.

Born in 1980 to hard-working impoverished parents in Newport News, Virginia, Michael Vick grew up in the Ridley public housing project during the Reagan years when funding for such projects was being largely withdrawn. The eastern part of Newport News where Vick grew up became so depressed that it was called 'Bad Newz,' the same name Vick gave to his kennel. As a very young boy Vick participated in a boys' club where he played basketball and baseball and, later, football. As a young player with the Warwick High School Raiders his abilities brought him to college attention and he accepted a scholarship offer from Virginia Tech, at which point his career took off. Within the white-sanctioned world of sport, however, black athleticism is celebrated because of its conceivable payback. It is through white ownership that disadvantaged youth, like Vick, assume social (market) value. By 2011, 100 percent of majority NFL owners, for instance, were (still) white and nearly 70 percent of players, black (Lapchick 2013).

While not a slave pen encircled by 15–20 foot walls, then, the Bad Newz community and Ridley housing projects were racialized enclosures from which Vick, through white ownership, was able to leave. The latter allowed him entry into a different kind of enclosed space, one valued for physical violence. Even here, though, Vick – like many black quarterbacks – would be criticized for not working hard enough (Broad 2013; Billings 2004).

Reducing Michael Vick's participation in dog fighting to moral depravity is therefore profoundly problematic in that it erases the historical and geographical forces of white settlement in the U.S. and the related contexts of power out of which all U.S. blood sports emerged (Nast 2013). Whereas in the past, Currier & Ives used the image of pit bull fighting to portray black men as incapable of being civilized, popular cultural portrayals of Vick have produced more of the same: in this case, an uppity football player whose imprisonment and bankruptcy are cast as his just desserts. Lost are opportunities to analyze how the Bad Newz ghetto itself and the ghettoized survivalist economies that the Reagan years bore are part and parcel of an uninterrogated supremacist logic that continues to move pitted fighting – and enclosures of all kinds – human and animal, licit and illicit – forward.

Notes

1 The ideas of race and breed co-evolved through various nineteenth-century discourses, including those related to supremacist endeavors. The suturing of the ideas was initially accomplished through the work of the amateur (and highly prolific) British scientist, Francis Galton (1822–1911) – half cousin of Charles Darwin

(1809–1882) – who coined the term 'eugenics' in 1883. His work was intended to show the applicability of Darwin's ideas about animal and plant evolution to the human context. He and many proponents of eugenics in the U.S., Britain, France, and Germany hence spoke interchangeably of race and breed. Indeed Darwin, himself, believed that breed and race were one and the same. In the U.S. their ideas would come to be embodied in the work of Madison Grant who worked to secure several restrictive immigration laws intended to secure the survival of the 'Nordic races,' his efforts receiving accolades from high-level governmental, scientific, and supremacist quarters, including the KKK. Thus, Spiro (2009: 233) writes that, "Thanks to Grant and scientific racism, the nation was now a refuge for the Nordics, where they could breed in peace, unmolested by alien strains." The interchangeability was rehearsed much earlier through other discursive means. See, for instance, 'Properties of different races of cattle' in Petzholdt (1846: 501) and Butler (1860) on various canine races and their attributes.

2 From part 3 of Michael Bur's unpublished work, *American Pit Bull Terriers Breed FAQ* (1995), widely circulated – often without attribution – on pit bull internet sites and blogs. See, for instance, http://stason.org/TULARC/animals/dogs/american-pit-bull-terriers-breed/03-History.html (viewed April 15, 2012).

3 Part 3 in Bur (1995).

4 Boars are omnivorous and live in a variety of ecological regions, including coastal plains, mountainous areas, forests, and forested bottomland areas. The impact of wild boar is great because they eat just about anything in their path: grass, nuts, acorns, berries, earthworms, roots, tubers, and insects as well as rodents, reptiles, amphibians, carrion, and small mammals.

5 Because there was no standardization at the time of slavery, there were no discrete breed names. Today, many hunters claim a variety of dogs to be ideal catch dogs, such as the Catahoula Bulldog, Cane Corso, and Dogo Argentino, the latter two recently being crossbred with fighting pit bull kinds so as to increase their size. Most of the literature on hog and cattle catch dogs circulates informally through the Internet or vanity presses. See, for instance, Kelley (2009) and Adele (2012). Note that these catch dogs are different from those used in sheepherding.

6 See www.bryantsreddevils.com/HogHunting.html.

7 See for instance, the following:

> 146: "The dog that I called a 'catch dog' was a bull terrier. They called him a catch dog."
> 209: "There was one catch dog, and the others were what they call plantation dogs. The catch dog was a sort of bull-dog, and the others were more of a hound."
> 256: "One or two in each pack ... were more of a bull-dog. They called them catch-dogs. ... Some of the men were bitten considerably."

8 Hardy County became divided during the Civil War, those allying themselves with Union forces withdrawing in 1866 to a separate portion that they called Grant County (Heishman 2013). It's not clear why this slave catcher would have been imprisoned in a Confederate prison, his slave catching suggesting that he would have been anti-Union.

9 Volume 1 of the FWP's Alabama Narratives, meanwhile, contains an entry called 'hoodooing de dogs' that speaks of how treed slaves could make themselves disappear (FWP 1936–1938b: 229–230).

10 *The Daily True Delta* was published daily (except Mondays) from 1849 to 1866 out of New Orleans, Louisiana. From 1849 to 1863, it was published by an Irishman, John Maginnis, who, at times, collaborated with other business partners. The fight notice was posted after the fact in the *New York Times* on May 24, 1856. John Maginnis's Irish roots may suggest that the dogs used in southern pit bull fighting came out of

'Irish' fighting-dog stock, these dogs first imported into Boston. This in contradistinction to the 'English' breeds that were apparently centered in New York.

11 See, for instance, Armitage who mentions, "Jim Powell, a colored fellow of New Kensington [Pennsylvania]" and the well known "Ben Merlette, a colored dog trainer from Montgomery, Alabama" (1935: 9, 23). According to the retired Oklahoma dogman, Randy Fox, Earl Tudor, the famous twentieth-century dog fighter and son of dirt-poor Kentucky farmers who started outbreeding 'English' with 'Irish' dogs in the 1930s, traveled in the 1950s to see an Arizona man who had begun breeding Tudor's line of dogs (Ed Richeson). He liked one of Richeson's dogs, and brought him back to Oklahoma. The man who cared for and trained the dog was African American (see http://rayfox6.tripod.com). Randy Fox noted that up through the 1960s African-American men rarely attended (white) dog fights. He approximated that if there were fifty or sixty dog fighters at an event, one or two might be African American (Randy Fox interview, December 12, 2011). In 1988, Georgia-born author Scott Ely wrote a novel *Pit Bull* set on a farm about to be foreclosed. The farmer decides to train his vicious pit bull to fight and hires an African-American fighting dog trainer. Randy Fox avers that this scenario is impossible: farmers would never have fought dogs; only "outlaws" did such things (Randy Fox interview, December 12, 2011). Randy Fox is a retired dogman in Oklahoma whose website is typical of many dogmen who write from firsthand experience about famous dogmen of their times and who provide informal histories of dog fighting in the past. In so doing, they often provide many details taken from archives and primary and secondary sources.

12 See, for instance, the several-line-long remark published in the May 4 paper (p. 5) under the heading: 'The Average Negro Girl' (1907). It reads: "Can be seen at every dog fight, card table and frolic—At every telephone where it is free, gossiping of the doings of the most common Negro class."

13 May 3, 1901. *Semi-Weekly Interior Journal* (Stanford, Kentucky).

Acknowledgments

Many thanks to the DePaul Humanities Center and DePaul University's Research Assistantship Program for their steady support. I am especially grateful to Anna Vaughn-Clissold, Jonathan Gross, and Peter Steeves. Thanks also to Kevin Doherty and Rachael Dimit for their extraordinary research assistance and to Merritt Clifton for informing me of the 1919 Vicksburg event. Lastly, my thanks to Rosemary and Katie for encouraging me to move forward with this project and for their incisive comments.

References

Adele, G 2012, *My wild backyard: Wild hogs*, Xlibris Corporation, www.Xlibris.com, Bloomington, IN.

Armitage, G 1935, *Thirty years with fighting dogs*, Jack Jones Press, Washington, DC.

Billings, AC 2004, 'Depicting the quarterback in black and white: A content analysis of college and professional football commentary,' *Howard Journal of Communications*, vol. 15 (4), pp. 201–210.

Broad, GM 2013, 'Vegans for Vick: Dogfighting, intersectional politics, and the limits of mainstream discourse,' *International Journal of Communication*, vol. 7, pp. 780–800.

Bur, M 1995, *American pit bull terriers breed FAQ* parts 1–3. Viewed April 20, 2014. www.faqs.org/faqs/dogs-faq/breeds/apbt/part1/#b; www.faqs.org/faqs/dogs-faq/breeds/apbt/part2/; www.faqs.org/faqs/dogs-faq/breeds/apbt/part3/.

Butler, F 1860, *Breeding, training, management, diseases, &c. of dogs*, Francis Butler, New York.

Cayton's Weekly 1919, 'The passing throng: race riots continue,' August 1, p. 2. Viewed September 16, 2014. http://chroniclingamerica.loc.gov/lccn/sn87093353/1919-08-02/ed-1/seq-1/.

Currier & Ives *c*.1884, *All broke up*, chromolithograph, issued as part of its 'Darktown comics' series, Library of Congress archives, Washington, DC. Viewed May 8, 2014. www.loc.gov/pictures/item/90708481/.

Currier & Ives *c*.1884, *A Sure Thing*, chromolithograph, issued as part of its 'Darktown comics' series, Library of Congress archives, Washington, DC. Viewed May 8, 2014. www.loc.gov/pictures/item/2002695766/.

Currier & Ives *c*.1884, *De white dog's got him!*, chromolithograph, issued as part of its 'Darktown comics' series, Library of Congress archives, Washington, DC. Viewed May 8, 2014. www.loc.gov/pictures/item/2002698813.

Ely, S 1988, *Pit Bull*, Grove Press, New York.

ESPN 2010, 'Tucker Carlson addresses Vick role'. Viewed April 20, 2014. http://sports.espn.go.com/nfl/news/story?id=5967015.

Federal Writers' Project (FWP) 1936–1938a, 'Bites dog's foot off,' *Born in slavery: Slave narratives from the Federal Writers' Project, 1936–38*, Arkansas Narratives, Volume II, Part 3. Library of Congress.

Federal Writers' Project (FWP) 1936–1938b, 'Hoodooin' de dogs,' *Born in slavery: Slave narratives from the Federal Writers' Project, 1936–38*, Alabama Narratives, Volume I, 229–230. Library of Congress.

Fox, R 2011, retired dogman, interview, December 12.

Fox, RK 1888, *The dog pit – or, how to select, breed, train, and manage fighting dogs, with points as to their care in health and disease*, Police Gazette, New York.

Gorant, J 2010, *The lost dogs: Michael Vick's dogs and their tale of rescue and redemption*, Gotham Books, New York.

Heishman, P 2013, 'Hardy county,' *e-WV: The West Virginia Encyclopedia*, May 31. Viewed April 19, 2014. www.wvencyclopedia.org/articles/238.

Humane Society of the United States (HSUS) 2011, 'Michael Vick and the HSUS call on feds to crack down on animal fighting spectators,' Press Releases, July 19. Viewed December 12, 2011. www.humanesociety.org/news/press_releases/2011/07/michael_vick_fighting_spectator_bill_071911.html; http://rayfox6.tripod.com.

Johnson, W 1999, *Soul by soul: Life inside the antebellum slave market*, Harvard University Press, Cambridge, MA.

Kelley, S 2009, *Hog hunting with dogs: The hogdoggers bible*, AuthorHouse, Bloomington, IN.

Lapchick, R 2013, 'The 2013 racial and gender report card of the National Football League,' *The Institute for Diversity and Ethics in Sport*, October 22. Viewed September 16, 2014. www.tidesport.org/RGRC/2013/2013_NFL_RGRC.pdf.

Massey, W 2012, 'Bloodsport and the Michael Vick dogfighting case: A critical cultural analysis,' MA Thesis, East Tennessee State University. Viewed September 16, 2014. http://dc.etsu.edu/etd/1513.

Morris, E 1979, *The rise of Theodore Roosevelt*, Coward, McCann & Geoghegan, New York.

Nast, HJ 2000, 'Mapping the "unconscious": Racism and the Oedipal family,' *Annals of the Association of American Geographers,* vol. 90 (2), pp. 215–255.

Nast, HJ 2013, 'Why dogs fight: Nineteenth-century coalmining, slavery, and the origins of the "pit",' Paper delivered at the Chicago Humanities Festival IIIV, November 9.

Nast, HJ forthcoming, *Petifilia*, University of Georgia Press, Athens, GA.

National Park Service (NPS) 2014, *History of the Andersonville Prison – Andersonville National Historic Site*. Viewed September 16, 2014. www.nps.gov/ande/historyculture/camp_sumter.htm.

Petzholdt, A 1846, 'Properties of different races of cattle,' *Lectures to farmers on agricultural chemistry*. Greeley & McElrath, New York, Tribune Buildings. Viewed September 16, 2014. https://archive.org/details/lecturestofarme00petzgoog.

Police Court 1901, *Alexandria Gazette and Virginia Advertiser*, March 19, p. 1.

Reckner, JR 2001, *Teddy Roosevelt's Great White Fleet*, Naval Institute Press, Annapolis, MD.

'Row at dog fight: Two Negros fatally and two others seriously wounded' 1904, *The Evening Bulletin*, Maysville, Kentucky, May 17, p. 1.

Schulte, B 2008, 'Saving Michael Vick's dogs: Pit bulls rescued from the football player's fighting ring show progress in an unprecedented rehabilitation effort,' *Washington Post*, July 7. Viewed September 16, 2014. www.washingtonpost.com/wp-dyn/content/story/2008/07/06/ST2008070602429.html.

Semi-Weekly Interior Journal 1901, May 3.

'Seven survivors of Michael Vick's dog fighting ring reunite five years later' 2012. Viewed April 25, 2014. www.lifewithdogs.tv/2012/11/seven-survivors-of-michael-vicks-dog-fighting-ring-reunite-five-years-later/.

Sieczkowski, C 2014, 'Michael Vick's former dogfighting pups will make you believe in happily ever after,' *The Huffington Post*, April 10. Viewed September 16, 2014. www.huffingtonpost.com/2014/04/10/michael-vick-dogs-vicktory_n_5119150.html.

Spiro, JP 2009, *Defending the master race: Conservation, eugenics, and the legacy of Madison Grant*, University of Vermont Press, Lebanon, VT.

Strouse, K with Dog Angel 2009, *Badd Newz: The untold story of the Michael Vick dog fighting case*, Dog Fighting Investigations Publications, LLC, Norfolk, VA.

The American Watch-Dog c1917, World War I postcard, Henry Heininger Co. Publishing, New York.

'The Average Negro Girl' 1907, *St. Louis Palladium*, p. 5.

The [New York] *Sun* 1898, no title, March 27, p. 3.

The Topeka Plaindealer 1910, 'Using the church for a bill board!,' September 9, p. 5.

U.S. House of Representatives 1868, *Trial of Henry Wirz, 2nd sess, 40th Congress*, Government Printing Office, Washington, DC. Viewed April 20, 2014. http://lccn.loc.gov/2011525460.

Watchful-Waiting c.1914, World War I postcard, Henry Heininger Co. Publishing, New York.

Works Progress Administration (WPA) 1938, *New Orleans city guide 1938*, Houghton Mifflin Company, Boston.

Part III

Hierarchies

9 Coyotes in the city

Gastro-ethical encounters in a more-than-human world

Gwendolyn Blue and Shelley Alexander

> One never eats entirely on one's own: this constitutes the rule underlying the statement 'one must eat well.' It is a rule offering infinite hospitality. And in all differences ... 'eating well' is at stake.
>
> (Jacques Derrida)

> In eating we are most inside the differential relationalities that make us who and what we are and that materialize what we must do if response and regard are to have any meaning personally and politically.
>
> (Donna Haraway)

Introduction

In February 2009, on an otherwise typical morning in a quiet residential Toronto neighborhood, a coyote jumped over a fence, grabbed a small dog named Zoe, and ran off into the woods. Zoe, who was never seen again, was not the first or last animal to be affected by this alleged culprit. That winter, several domestic cats had gone missing, and other dogs had been bitten. Residents reported seeing coyotes displaying "unusually bold and aggressive behavior" in the immediate area (Reilly 2009). This behavior was primarily attributed, by experts and residents alike, to people advertently or inadvertently providing food for coyotes and, in turn, habituating them to the presence of humans (Johns 2009).

Some local residents and subsequently the media began referring to the coyote as Neville in reference to the street on which Zoe's abduction took place. Neville Park Boulevard is located in a picturesque area of the city known as the Beaches (or the Beach), with a topography characterized by gullies and ravines adjacent to Lake Ontario. The natural setting of the region is a draw for residents and visitors, human and otherwise.

In response to this incident, the City of Toronto's Animal Services deployed several tactics to encourage Neville and his kin to leave the area, including traps as well as hazing mechanisms such as paintballs and noisemakers. When these failed, the City declared that the only remaining option was to trap and euthanize the animal (Spears 2009). Across Canada, coyotes are routinely sentenced to death when such conflicts with humans arise (Alexander and Quinn 2011).

Due to their legal designation as a pest species, coyotes are not offered protection by the state.

This particular verdict proved to be politically divisive (Bonoguore 2009). Some residents were emphatic that problem coyotes should be removed in order to make the neighborhood safe for domestic animals and children. Others took the position that Neville should be spared, given that coyotes are an integral part of the ecosystem. On March 29, a local resident circulated a "Save Neville" petition (Morrow 2009). Taking up the cause, the Humane Society of Canada posted the petition on its website, as well as requested in writing that the City and the Minister of Natural Resources pursue alternative measures. The following day the City commuted Neville's sentence: once trapped, the coyote would be relocated to a safe area outside of city limits (Hammer 2009).

While unique in some regards, Neville's story is indicative of human–coyote encounters taking shape in major metropolitan areas across North America. Over the past decade, sightings of coyotes and perceived human–coyote conflicts have become increasingly common in urban centers such as Toronto, Vancouver, Edmonton, Calgary, Los Angeles, Chicago, Washington, and New York. Typically wary of humans, coyotes tend to keep their distance. But they can lose their fear and become aggressive if they grow accustomed to human food sources. The connection between diet (i.e., anthropogenic food sources) and reported conflicts indicates that without adequate management, serious human–coyote conflicts can emerge. Increasingly, efforts to promote coexistence with coyotes focus on the importance of managing potential food sources and re-instilling fear through aversive conditioning (i.e., hazing).

Approached through an ethical rather than managerial or instrumental lens, this analysis suggests that more is at stake in learning to coexist with urban coyotes than managing food sources or frightening coyotes. As with other non-domesticated carnivores such as cougars, wolves, and bears, urban coyotes provide a test case for our collective capacity to live with diversity and difference in more-than-human contexts (Beatley and Bekoff 2012; Plumwood 2002). Drawing on recent work in human–animal studies and more-than-human geographies, we examine what it means to live responsibly and ethically with nondomesticated carnivores. Nourished by Haraway's companion species encounters as well as Derrida's gastronomic hospitality, we outline how ethical engagements necessarily involve grappling with how to "eat well" in multispecies contexts. As the epigraph to this chapter illustrates, eating well is an acknowledgment that we never eat entirely on our own (Derrida 1991), and that we invariably eat in complex multispecies worlds (Haraway 2003, 2008). Situating ethics in dynamic and diverse ecologies of consumption, we call into question deeply rooted assumptions about human subjectivity and exceptionalism as well as attendant notions of moral purity.

What we call gastro-ethical encounters is a contextual, place-based, pragmatic, and inherently messy approach to ethics. Never pure, it involves acknowledging three inescapable conditions: that humans need to eat as do other living creatures; that we eat in multispecies environments comprised of different entities with conflicting tastes and bodily requirements; and that eating and killing are intimately entangled, as are life and death. In articulating this account of ethics, we take direction from Haraway with respect to the potential of thinking with narrative figures. Whereas Haraway (2008) takes her dog Cayenne Pepper as her central figure, we engage Neville the coyote. Methodologically, Neville's saga provides not so much a case study from which to draw generalizations, but a specific "attachment site" (Haraway 2010: 54) from which to grapple with ethico-political unfoldings in urban environments.

In what follows, we first clarify what it means to position Neville and urban coyotes more generally as companion species. This relational ontological approach has significant implications for how we approach ethics. The subsequent section discusses in more detail the phenomena of urban coyotes across North America. Next, drawing on recent work in animal geography, we address the challenges emerging in metropolitan centers as the coyote draws the urban into its matrix of livable habitat. In turn, raising the question of the animal as posed by Haraway and Derrida, we sketch an ethical framework that attends to the complexity of living with predators, large and small. To conclude, we reflect on the political possibilities that urban coyotes like Neville hold for a convivial if not fraught politics of coexistence.

Coyotes as companion species

To begin, we offer a qualification about classifications and, in turn, about ethics. One of the most persecuted indigenous carnivores in North America, coyotes can be addressed and understood in many ways. *Canis latrans*. Song dog. Prairie wolf. Pest. Trickster. Neville. These names serve as classifications, operating as powerful ordering devices that frame relations in particular ways. Never neutral, classifications provide significant sites of political and ethical engagement (Bowker and Star 2000). Classifications reveal and highlight aspects of the material world, but they also conceal and obscure others. Classifications, in short, matter, warranting measured consideration from the outset.

We approach coyotes, both as individuals and as populations, as companion species, to borrow Haraway's (2003, 2008) provocative abstraction. Haraway defines companion species as a material-semiotic concept, intended not so much a fixed category but as "a pointer to an ongoing 'becoming with'" (2008: 16). Companion species encompass but also extend beyond companion animals, the domesticated creatures with whom we share our lives. Rooted in a relational, non-dualist ontology (Braun 2008), companion species serve as an important reminder of the "the implosion of nature and culture" in the lives of creatures who are "bonded in significant otherness" wherein "co-constitution, finitude, impurity, historicity and complexity" are all that is (Haraway 2003: 16). Significantly, companion species signal the ways in which collective worlds are made in specific spaces and by specific creatures whose identities and characteristics do not preexist their ongoing relations (cf. Barad 2007). Deployed with increasing frequency in animal and more-than-human geographies, companion species address the collective becoming of human and nonhuman individuals and populations, who are simultaneously bonded through histories of cohabitation but who are also different from one another in significant ways (Bingham 2006; Collard 2012; Dempsey 2010; Hinchliffe 2010; Lorimer 2010; Notzke 2013).

Haraway's relational and embodied account of multispecies encounters offers a dramatically different approach to ethics than is espoused by rights-based or utilitarian discourses. Grounded in a dualist rather than a relational ontology, these latter formulations tend to assume that moral agents are bounded, atomistic, and autonomous individuals who are inherently separate from one another as well as from the environments in which they are located (for detailed critiques of dualist ontologies in ethics, see Ingold 2000; Smith 2001; Whatmore 2002). Shifting the focus of ethics away from the singular, decontextualized subject, Haraway draws attention to the situated relations among individuals and groups. This ontological position resonates with geographers who, in various ways, articulate relational ethics through an attention to place and context (e.g., Lynn 1998) as well as multispecies difference (e.g., Bingham 2006; Whatmore 2002).

In line with these efforts, we advance a relational approach to ethics that takes corporeality, context, and difference seriously and that does not presume or reinforce anthropocentric, individualistic assumptions. In doing so, we move Haraway's companion species, and its attendant ethics, beyond benign encounters among mortal creatures. Rather, our concern lies with those who by necessity, proclivity, or desire, consume the flesh of other beings.

Positioned as companion species, coyotes are not static entities that exist 'out there' in a state of nature separate from human interaction. Nor is our existing knowledge about coyotes complete such that we can point to maps, models, and scientific descriptions as devices of prediction and control. Adaptable, variable, and intelligent, coyotes confound existing expectations and assumptions, refusing to remain contained by the pervasive dualisms that organize Euro-American thought and practice. Indeed, coyotes' reputation as a trickster is well earned, and informs our own politics and ethics (cf. Haraway 1989; Phelan 1996).

Urban coyotes

Unlike other carnivores in North America, such as wolves and bears for whom centuries of persecution and habitat destruction have reduced their historical range, coyotes have adapted to human presence and to humanized landscapes. Some report that coyotes were ubiquitous across the open plains or short-grass prairies of western North America until the eighteenth century (Fox 2006; Parker 1995). It is speculated that the arrival of European settlers opened up additional niches for coyotes. Coyotes can now be found from California to Newfoundland, and from Panama to Alaska, inhabiting every province in Canada and every state in the U.S., with the exception of Hawaii (Bekoff and Gese 2003). Coyotes also inhabit major metropolitan centers across North America. The presence of coyotes in urban centers has been attributed to habitat destruction and the availability of food and water in urban environments (Fox and Papouchis 2005; Garthwaite 2012).

Part of the coyote's adaptability to human presence lies with its behavioral plasticity particularly with respect to eating. Coyotes have been shown to consume a range of food including rodents, birds, fruit, insects, even family pets (Lukasik and Alexander 2012). Coyotes also tend to keep a low profile and avoid humans, even in highly urbanized contexts such as Chicago (Gehrt 2004; Gehrt *et al.* 2009). Many people in cities often do not know that they live in the presence of coyotes, and can be surprised when they encounter or observe one.

Awareness of the "hidden masses" of these animals tends to be raised when encounters and conflicts occur (Beatley and Bekoff 2012). Importantly, the vast majority of coyote–human interactions in urban settings are benign, limited primarily to sightings (Alexander and Quinn 2011; Lukasik and Alexander 2011). As we will explore in more detail later, people are perhaps more likely to report sightings of coyotes if they believe these creatures are 'out of place' in urban contexts. Coyotes do attack small dogs and cats; only in rare instances do they attack humans (Bounds and Shaw 1994; Timm and Baker 2007; White and Gehrt 2009). While statistically the number of people and domestic animals bitten by coyotes is low, these encounters often serve to reinforce fear and prejudice as well as to amplify perceptions of risk (Alexander and Quinn 2012; Beatley and Bekoff 2012; Fox 2006). Whether experienced first- or secondhand, coyote incidents can have a strong influence on people's perceptions, particularly if loved ones (domestic pets or humans) are injured in an encounter. Most often, these encounters prove fatal for coyotes.

When serious conflicts do arise, they are typically linked with food. Intentional or unintentional feeding can lead to habituation, where coyotes lose their fear of humans (Alexander and Quinn 2011; Fox 2006; Lukasik and Alexander 2011). Unsecured garbage, pet food, free-roaming cats and small dogs, fruit trees, even ornamental plants are among the many comestibles that attract coyotes.

Coyotes also present a possible risk to the health of humans and domestic canines as they can serve as a vector for diseases that are also carried by domestic dogs. This creates concern among some veterinarians and public health officials as people and dogs could increasingly interact with coyotes in urban environments. Unsubstantiated media reports of rabies, mange and other canine pathogens often serve to reinforce and exacerbate preexisting fears (Alexander and Quinn 2012). Despite the fact that any canid can serve as a host, some veterinarians and public health officials have targeted coyotes as a risk due to their role as a host to the parasite *Echinococcus multilocularis* (Catalano *et al.* 2012). This parasite can be transferred from domestic to wild canids, or vice versa, presenting a risk to humans acquiring the parasite through contact with domestic dogs. If humans are infected the consequences can be severe, prompting some officials to promote increased surveillance and in some cases removal of urban coyote populations.

The first response to coyote conflicts tends to involve removal of the animal by trapping, poisoning, or extermination. Such efforts have proven ineffective, however, as they do not address the root cause of the problem: access to human-sourced food (Alexander and Quinn 2012; Fox 2006). When a coyote is removed from an area, the void is usually filled by a coyote from surrounding populations. If culls (large-scale, state-sanctioned killing) are enacted, this can serve to increase reproductive and recruitment rates, leading to higher pre-cull populations overall (Berger 2006).

Many biologists, wildlife managers and animal rights advocates argue that it is necessary instead to address the underlying problem through public education, namely instructions on how to manage potential food sources to avoid attracting coyotes, as well as lessons in hazing coyotes to scare them away from populated regions (Beatley and Bekoff 2012; Fox 2006). Organizations such as Project Coyote (USA), Coyote Watch Canada, the Beach Coyote Coalition of Toronto and the Stanley Park Ecological Society of Vancouver, to list just a few, have been instrumental in raising public awareness of coexistence through education and outreach. Municipalities across North America have also played a central role in promoting coexistence with coyotes. To cite but one example, the City of Vancouver has implemented a successful coyote coexistence program that includes proactive public education and outreach, bylaws against wildlife feeding, as well as a coyote hotline that facilitates monitoring (Worcester and Boelens 2007).

While such management initiatives are an important response to urban coyotes, they are insufficient on their own. A potentially more difficult task, one for which critical animal geographers are well positioned, entails calling into question and reconfiguring deeply rooted notions about ethics and responsibility as well as attendant concepts such as belonging and otherness. It is to these issues that we now turn our analytic lens.

Urban coyotes: challenges to *anima urbis*

Animal geographers and urban political ecologists have long questioned the wide-spread assumption that cities are the exclusive domain of human activity (Braun 2005; Heynan 2013; Hinchliffe *et al.* 2005; Lynn and Sheppard 2004; Wolch 1998, 2002; Wolch *et al.* 1995; Philo 1995). Disrupting the conceptual and material separation between wilderness and culture as well as country and city that shape contemporary urban theory and practice, these scholars seek to re-wild urban spaces by situating cities as more-than-human, rather than strictly human, sites of interaction and engagement. As Wolch (2002) puts it, this re-wilding means envisioning a new form of urbanism, an *anima urbis*, which attempts to "bring animals and plants back in" (Wolch and Emel 1995). This move is vital not only in terms of developing a better understanding of cities, but also of engaging in a politics premised on ethical relations with nonhuman creatures.

Urban coyotes present a compelling example of the challenges of *anima urbis* particularly as an entrenched geographical imaginary founded on maintaining material and symbolic boundaries between nature and culture breaks down. Urban coyotes inhabit areas that have been historically considered to be the domain of humans. As such, their presence disrupts norms and conventions that have emerged over the past two centuries, particularly those associated with the killing of animals. A significant component of the organization of urban spaces and identities hinges on the allocation of food production to rural areas and food consumption to urban ones (Gaynor 1999; Philo 1995; Philo and Wilbert 2000). This has entailed the removal of certain animals whereby productive animals are relegated to rural spaces and wild animals to the wilderness. Over time, activities associated with the killing of animals for food such as meat markets and slaughterhouses were also excised from the city, reinforcing urban sensibilities to be increasingly defined in opposition to rural ones (Philo 1995). As Gaynor (1999) observes, this process of spatial reordering classified certain animals as belonging to the city and relegated others as out of place. Household pets, in particular, were considered an integral (and normal) component of middle-class urban lifestyles of consumption. Philo and Wilbert (2000: 11) describe the spatial ordering as follows:

> Zones of human settlement ('the city') are envisaged as the province of pets or 'companion animals' (such as cats and dogs), zones of agricultural activity ('the country side') are envisaged as the province of 'livestock' animals (such as sheep and cows), and zones of unoccupied lands beyond the margins so settlement and agriculture ('the wilderness') are envisaged as the province of wild animals (such as wolves and lions).

By inhabiting urban settings, coyotes refuse to remain within such tidy geographical orderings and imaginaries. In transgressing these categories, coyotes can also be viewed as out of place (Cresswell 1996) and risk inciting potentially dangerous reactions from the human community.

Rather than strictly relegating nonhuman animals to human-generated spatial orderings, recent work in animal geography has recognized nonhuman agency in shaping collective understandings of space and place. The scholarship that is most germane to our discussion highlights the important role that carnivores play in creating collective space as well as the challenges of sharing space with these animals (Collard 2012; Dempsey 2010; Gullo *et al.* 1998). For example, Gullo *et al.* (1998) trace changing attitudes of humans towards cougars in Los Angeles, as well as the ways in which cougar attitudes towards humans also changed. This capacity for mutual learning can provide fruitful avenues for coexistence:

> If people can learn that life in the human–animal borderlands implies a duty to know their cougar neighbours ... and if cougars can learn that borderlands survival means keeping a respectful distance, maybe communities of the urban-wildlands fringe can once again rest safe.
>
> (1998: 158)

Coexistence, in this sense, requires maintaining and marking a certain distance between human and cougar populations. Such thinking is the premise behind coyote hazing programs that have been successfully implemented by groups like Coyote Watch Canada.

Similarly, Collard (2012) examines the ways in which safe spaces on Vancouver Island are made and unmade by both cougars and humans. Her analysis also concludes by emphasizing the need to keep a respective distance between humans and wild carnivores. Herein, she raises an important question with regard to Haraway's positioning of ethics. Citing the claim that "we are in a knot of species co-creating each other [where] response and respect are possible only in those knots, with actual animals and people looking back at each other" (Haraway 2008: 42), Collard aptly notes that an ethics of intimacy works only in relation to certain animals, such as companion animals, where there is little to no threat of predation. Concluding that Haraway is only "partly right" when it comes to ethical relations with large carnivores, Collard argues that the challenge in being responsible to these creatures lies with "recognizing the lives of those we do not always see but with whom we are immeasurably entangled" (Collard 2012: 38).

Importantly, this study points to the limitations of ethical frameworks that are predicated on visibility as well as on the possibility of convivial face-to-face encounters with non-domesticated carnivores. Extending these insights but drawing on a slightly different interpretation of companion species, we locate the promise and possibility of an ethics of coexistence in eating, one of the most intimate, mundane, inescapable, and invariably violent dimensions of life.

Gastro-ethical encounters

Haraway's description of companion species is based on two registers, visual and alimentary, although it is the former that tends to be emphasized in her work and elsewhere. The majority of Haraway's reflections, in *Companion Species Manifesto* (2003) as well as *When Species Meet* (2008), hinge on an approach to ethics grounded in *respecere*, the capacity to look and look back (2008: 164). In defining companion species, Haraway draws on the Latin *specere* (to look and to behold) as well as the definition of species in logic as a "mental impression or idea" to signal the intimate link between thinking and seeing (2008: 17).

The other dimension of companion species derives from the Latin *cum panis* (with bread), in tandem with the verb, "to companion," meaning "to consort, to keep company" (2008: 17). In the final chapter of *When Species Meet*, Haraway emphasizes that it is through the mundane, quotidian practices of producing, procuring, distributing, and consuming food that we contribute to the ongoing processes that constitute identities and collective worlds. Moreover, it is in the practices of eating and the relations these practices engender that she locates ethical directives.

Specifically, Haraway evokes Derrida's (1991) understanding of Other-ethics, phrased in terms of how to "eat well." To provide a brief background, in his interview with Jean-Luc Nancy published as '"Eating Well" or the Calculation of the Subject,' Derrida calls into question the entrenched binary opposition between the human and the animal that has long informed Western philosophical traditions. While Derrida's (1991, 2002) general concern lies with deconstructing the implicit anthropocentrism in Heideggerian and post-Heideggarian thought, in this particular interview he critiques the work of one of the foremost ethical theorists in the tradition of continental philosophy: Emmanuel Levinas. While it is beyond the scope of this chapter to provide a detailed and nuanced description of this critique (Calarco 2004 provides an instructive overview), Derrida's general argument is that Levinas, in attempting to overturn classical humanist approaches to ethics, serves ultimately to reinforce them. What Levinas fails to do, according to Derrida, is to "sacrifice sacrifice," by which he means that this approach stops short of questioning the noncriminal putting to death of nonhuman life. The place of animals in Levinas's thought echoes the place of animals within the dominant schema of Judeo-Christian thought, where "thou shalt not kill" applies only to human life. For Derrida, if we can no longer support the oppositional lines between human and animal, then our ethical approaches must alter accordingly. Eating well is the direction Derrida proposes, a situated approach to ethics that involves "determining the best, most respectful, most grateful and also most giving way of relating to the other and of relating the other to the self" (1991: 114).

For Haraway, eating well offers two important insights. First, it is an acknowledgement that we never eat entirely on our own, signaling the diversity of bodies and entities that are evoked in multispecies environments. Second, eating well provides an opening for "infinite hospitality" (Derrida 1991: 109). Insofar as it locates ethical inquiry in the ongoing unfolding of the world, eating well offers possibilities for immanent re-worlding, as Haraway puts it.

Taking Derrida's insights further, Haraway explains that eating well offers a grounded engagement for creatures entangled in a world in which eating is inextricably linked to killing. Haraway contends that attempts to reconcile eating well with the imperative "thou shalt not kill" amounts to a philosophical and political misstep, one that is predicated on "forgetting the ecologies of all mortal beings who live in and through the use of one another's bodies" (2008: 79). Grounding ethics firmly in worldly relations means acknowledging the following: "There is no way to eat and not to kill, no way to eat and not to become with other mortal beings to whom we are accountable, no way to pretend innocence and transcendence or a final peace" (2008: 295). Killing may be direct or indirect, acknowledged or denied, but it is ever-present in eating as it is in the multispecies unfoldings of the world. To live by the imperative "thou shalt not kill" means necessarily marking a certain portion of living creatures as other such that the imperative does not apply. To evoke the terminology of Agamben (1998), this serves to confine certain living creatures to the realm of *zoe* or "bare life," to that which is killable, rather than situating them in the realm of *bios*, to those with political, individual, and historical lives.

The alternative, Haraway suggests, is to aim for a more modest and responsive position: thou shalt not make killable. This marks a refusal to position certain living creatures as other to guiding ethical imperatives. This is not an admonition that "nature is red in tooth and claw" (2008: 79). Nor does it mean that any and all kinds of eating are acceptable. Rather, it offers an acknowledgement of ecological relations in which heterogeneous beings are invariably immersed in webs of consumption where eating and killing are never separate.

While Haraway's insights open novel ethical territory, for the most part, her work skirts the alimentary entanglements of coexistence that companion species otherwise invites us to consider. With the exception of the concluding chapter, *When Species Meets* focuses on companion animals (Haraway's dog Cayenne Pepper) and humans (her father Frank and colleague Rusten). She does not regard these creatures as food, nor do they regard her as food. As she emphatically states, "one does not eat one's companion animals (nor get eaten by them); and one has a hard time shaking colonialist, ethnocentric, ahistorical attitudes towards those who do (eat or get eaten)." (2003: 14) Her concerns focus primarily on those creatures "who cannot and must not assimilate one another but who must learn to eat well, or at least well enough that care, respect and difference can flourish in the open." (2008: 287).

Our approach, by contrast, addresses the challenges of coexisting with non-domesticated carnivores. As landscapes become increasingly urbanized, and as coyotes inhabit remnant habitat in urban spaces, we are forced to confront difficult questions. What does it mean, ultimately, to live and eat well among coyotes given that they might harm our companion animals or our children? What practices and relations are demanded of coyotes, humans, and domestic animals in the process of re-wilding urban spaces? Which lives will be spared, and which will be sacrificed? While offering no easy solutions, gastro-ethics provides a foundational ground for responsive and response-able relations among more-than-human companions.

Gastro-ethics for a convivial *anima urbis*

To summarize, this analysis takes us to the heart (or, perhaps more aptly, the gastrointestinal tract) of a pressing challenge in more-than-human urban environments: coexistence with non-domesticated carnivores. We argue that coexistence is best worked out through gastro-ethical encounters with an attendant imperative to eat well. Eating, in this regard, is not a strictly human-centered act but is a relational, multispecies engagement that serves to co-constitute bodies, environments, and worlds. Living creatures work out gastro-ethical commitments in ways that are necessarily untidy, never pure, sometimes bloody, always contingent, and often opportunistic.

On one level, eating well in the presence of urban coyotes can be remarkably straightforward, requiring easily implemented behavioral changes such as securing garbage, keeping children and small domestic animals close at hand, as well as monitoring the accessibility of yard plants. On another level, the challenge is more taxing. Instrumental solutions, such as the management of behavior, risk neglecting important questions concerning how to get along, ethically and responsibly, in the face of multispecies difference.

Two implications follow from the approach to ethics outlined in this analysis. Situated in relational ontology, corporeality, and multispecies difference, gastro-ethics challenges the assumption that ethical engagements can be worked out in ways that are removed from the quandaries of embodied life. Embodiment is not so much a matter of being situated or embedded in a preexisting world. Instead, it is an acknowledgment that, like other creatures, we are of the world. As such, ethics is not about finding the "right or proper response to a radically exterior/ized other"; rather, it involves being responsible for and to "the lively relationalities of becoming of which we are a part" (Barad 2007, cited in Haraway 2008: 289). Gastro-ethics acknowledges that we are always becoming with other individuals and species, many of whom have radically different tastes and sensibilities than our own.

Second, rather than assume that ethical encounters are rooted in the possibility of retaining the integrity of discrete bodies, gastro-ethics holds from the outset that the alimentary ties that bind companion species are not readily severed. To eat is to assimilate and hence to kill. As such, mortal and moral creatures are invariably incapable of fulfilling the Judeo-Christian dictate: thou shalt not kill. To hold this imperative and to simultaneously eat means that we must mark certain bodies as other. We must also forget, momentarily, our inherent animality and ecological identity (Plumwood 2002). This forgetting in turn underpins and motivates human exceptionalism. By contrast, in recognizing that eating necessarily involves killing, we open ourselves to the capacity to be responsive and response-able to the plight of all living beings who invariably harm others by virtue of the need to eat.

In foregrounding the links between eating and killing, gastro-ethics moves critical animal geography into potentially new ethico-political territory in which disrupting the primacy of the human is the central concern (cf. Anderson 2014). Guided by the explorations of more-than-human geographies (e.g., Bingham 2006; Whatmore 2002), we use the alimentary entanglements that bind companion species as a starting point to position urban environments as comprised of a motley crew of characters who are intimately connected with one another through eating and, by extension, through killing. Agency and culpability are hence dispersed, as coyotes, humans, dogs, mice, deer, even parasites shape and reshape collective worlds through webs of consumption where flourishing is partial, never total. At any moment, some will live and some will die in Earth's never-ending multispecies encounters. It is in the "generative indigestion" (Haraway 2010: 53) that emerges in the contact zones of multispecies encounters that we locate political and ethical possibility.

Parting bites

This foray into "the obligations of emergent worlds where untidy species meet" (Haraway 2008: 294) has been guided by a specific material-semiotic companion: Neville. We devote our final words to his continuing saga.

Among all the coyote conflicts reported in Canadian print media from 1998 to 2010 (Alexander and Quinn 2012), Neville's story is unique. Neville was the only coyote over this time period to be publicly granted a stay of execution. Moreover, Neville was the only coyote given a name. This classification marked him not as a disposable other but as a unique denizen who belonged in, of, and to the neighborhood he inhabits, moving him in essence from *zoe* to *bios*. Although Neville allegedly killed cherished domestic animals, his life was ultimately spared, due to the efforts of a broader community who took responsibility for his welfare. At the time of writing and to the best of our knowledge, Neville is alive and free, having eluded all attempts at capture. This hints at the potential for eating well, a convivial politics of multispecies difference in urban contexts.

While the past has not laid much ground for optimism, understanding the city to be habitat and home for non-domesticated carnivores such as Neville might enable engagements with urban environments as ethical spaces in ways not previously considered. In reminding us that all species, including our own and our domestic kin, are part of and not separate from the food chain, Neville ultimately decenters human subjectivity in ways that Western theoretical traditions have yet to adequately address. As such, Neville provides a grounded, realistic, fecund, if not uncomfortable, starting point for ethical engagement and political possibility.

References

Agamben, G 1998, *Homo sacer: Sovereign power and bare life*, trans. D Heller-Roazen, Stanford University Press, Stanford.

Alexander, S and Quinn, M 2011, 'Coyote (*Canis latrans*) interactions with humans and pets reported in the Canadian print media (1995–2010),' *Human Dimensions of Wildlife*, vol. 16 (5), pp. 345–359.

Alexander, S and Quinn, M 2012, 'Portrayal of interactions between humans and coyotes (*Canis latrans*): Content analysis of Canadian print media (1998–2010),' *Cities and the Environment (CATE). Special Topic Issue: Urban Predators*, vol. 4 (11). Viewed September 16, 2014. http://digitalcommons.lmu.edu/cate/vol4/iss1/9/.

Anderson, K 2014, 'Mind over matter: On decentering the human in Human Geography,' *Cultural Geographies* vol. 21 (1), pp. 3–18.

Barad, K 2007, *Meeting the universe halfway: Quantum physics and the entanglement of matter and meaning*, Duke University Press, Durham, NC.

Beatley, T and Bekoff, M 2012, 'City planning and animals: Expanding our urban compassion footprint,' in C Basta and S Moroni (eds), *Ethics, design and planning of the built environment*, Springer, New York.

Bekoff, M and Gese, E 2003, 'Coyote (*Canis latrans*),' *USDA National Wildlife Research Center – Staff Publications*, Paper 224.

Berger, K 2006, 'Carnivore-livestock conflicts: Effects of subsidized predator control and economic correlates on the sheep industry,' *Conservation Biology*, vol. 20 (3), pp. 751–761.

Bingham, N 2006, 'Bees, butterflies and bacteria: biotechnology and the politics of nonhuman friendship,' *Environment and Planning A*, vol. 38 (3), pp. 483–498.

Bonoguore, T 2009, 'Some residents want to save the coyote in their midst,' *The Globe and Mail*, March 30, p. A8.

Bounds, D and Shaw, W 1994, 'Managing coyotes in U.S. National Parks: Human-coyote interactions,' *Natural Areas Journal* vol. 14 (4), pp. 280–284.

Bowker, G and Star, S 2000, *Sorting things out: Classification and its consequences*, MIT Press, Cambridge, MA.

Braun, B 2005, 'Environmental issues: Writing a more-than-human urban geography,' *Progress in Human Geography*, vol. 29 (5), pp. 635–650.

Braun, B 2008, 'Environmental issues: Inventive life,' *Progress in Human Geography*, vol. 32 (5), pp. 667–679.

Calarco, M 2004, 'Deconstruction is not vegetarianism: Humanism, subjectivity, and animal ethics,' *Continental Philosophy Review*, vol. 37 (2), pp. 175–201.

Catalano, S, Manigandan, L, and Massolo, A 2012, '*Echinococcus multilocularis* in urban coyotes, Alberta, Canada,' *Emerging Infectious Disease*, vol. 18 (10), pp. 1625–1628.

Collard, R-C 2012, 'Cougar–human entanglements and the biopolitical un/making of safe space,' *Environment and Planning D: Society and space*, vol. 30 (1), pp. 23–42.

Cresswell, T 1996, *In place, out of place: Geography, ideology and transgression*, University of Minnesota Press, Minneapolis.

Dempsey, J 2010, 'Tracking grizzly bears in British Columbia's environmental politics,' *Environment and Planning A*, vol. 42 (5), pp. 1138–1156.

Derrida, J 1991, '"Eating well" or the calculation of the subject,' in E Cadava, P Connor, and JN Nancy (eds), *Who Comes After the Subject?*, Routledge, New York and London.

Derrida, J 2002, 'The animal that therefore I am (more to follow),' *Critical Inquiry*, vol. 28 (2), pp. 369–418.

Fox, C 2006, 'Coyotes and humans: Can we coexist?,' *Proceedings of Vertebrate Pest Conference* March 6–9, Vertebrate Pest Council, University of California Davis, pp. 287–293.

Fox, C and Papouchis, C 2005, *Coyotes in our midst: Coexisting with an adaptable and resilient carnivore*, Animal Protection Institute, Sacramento, CA.

Garthwaite, J 2012, 'Learning to live with urban coyotes,' *New York Times*, October 24. Viewed November 15, 2013. http://green.blogs.nytimes.com/2012/10/24/learning-to-live-with-urban-coyotes/?_php=true&_type=blogs&_r=1&.

Gaynor, A 1999, 'Regulation, resistance and the residential area: the keeping of productive animals in twentieth-century Perth, Western Australia,' *Urban Policy and Research*, vol. 17 (1), pp. 7–16.

Gehrt, S 2004, 'Ecology and management of striped skunks, raccoons, and coyotes in urban landscapes,' in N Fascione, A Delach, and ME Smith (eds), *People and predators: From conflict to coexistence*, Island Press, Washington, DC.

Gehrt, S, Anchor, C, and White, L 2009, 'Home range and landscape use of coyotes in a metropolitan landscape: conflict or coexistence?,' *Journal of Mammalogy*, vol. 90 (5), pp. 1045–1057.

Gullo, A, Lassiter, U, and Wolch, J 1998, 'The cougar's tale,' in J Wolch and J Emel (eds), *Animal geographies: Place, politics and identity in the nature-culture borderlands*, Verso, London.

Hammer, K 2009, 'Killer coyote dodges death sentence,' *The Globe and Mail*, March 31, p. A11.

Haraway, D 1989, *Simians, cyborgs and women*, Routledge, London.

Haraway, D 2003, *The companion species manifesto: Dogs, people and significant otherness*, Prickly Paradigm Press, Chicago.

Haraway, D 2008, *When species meet*, University of Minnesota Press, Minneapolis and London.

Haraway, D 2010, 'When species meet: Staying with the trouble,' *Environment and Planning D*, vol. 28 (1), pp. 53–55.

Heynan, N 2013, 'Urban political ecology: The urban century,' *Progress in Human Geography*, doi: 10.1177/0309132513500443.

Hinchliffe, S 2010, 'Where species meet,' *Environment and Planning D: Society and Space*, vol. 28 (1), pp. 34–35.

Hinchliffe, S, Kearnes, M, and Degen, M 2005, 'Urban wild things: A cosmopolitical experiment,' *Environment and Planning D: Society and Space*, vol. 23 (5), pp. 643–658.

Ingold, T 2000, *The perception of the environment: Essays on livelihood, dwelling and skill*, Routledge, London.

Johns, E 2009, ' "Coyotes" future fuzzy: May be sent to sanctuary or euthanized to protect public safety,' *The Hamilton Spectator*, February 28, p. A3.

Lorimer, J 2010, 'Elephants as companion species: the lively biogeographies of Asian elephant conservation in Sri Lanka,' *Transactions of the Institute of British Geographers*, vol. 35 (4), pp. 91–506.

Lukasik, V and Alexander, S 2011, 'Human-coyote interactions in Calgary, Alberta,' *Human Dimensions of Wildlife*, vol. 16 (2), pp. 114–127.

Lukasik, V and Alexander, S 2012, 'Spatial and temporal variation of coyote (*Canis latrans*) diet in Calgary, Alberta,' *Cities and the Environment, Special Topic: Urban Predators*, vol. 4 (11). Viewed September 16, 2014. http://digitalcommons.lmu.edu/cgi/viewcontent.cgi?article=1097&context=cate.

Lynn, W 1998, 'Animals, ethics and geography,' In J Wolch and J Emel (eds), *Animal geographies: Place, politics and identity in the nature-culture borderlands*, Verso, London.

Lynn, W and Sheppard, E 2004, 'Cities,' in S Harrison, S Pile, and N Thrift (eds), *Patterned ground: Entanglements of nature and culture*, Reaktion, London.

Morrow, A 2009, 'Save-the-coyote petition urges city to abandon attempt to destroy the animal,' *Toronto Star*, March 29, p. A4. Viewed September 16, 2014. www.thestar.com/news/2009/03/29/savethecoyote_petition_urges_city_to_abandon_attempt_to_destroy_the_animal.html.

Notzke, C 2013, 'An exploration into political ecology and nonhuman agency: The case of the wild horse in western Canada,' *The Canadian Geographer*, vol. 57 (4), pp. 389–412.

Parker, G 1995, *Eastern coyote: The story of its success*, Nimbus Publishing, Halifax, NS.

Phelan, S 1996, 'Coyote politics: Trickster tales and feminist futures,' *Hypatia*, vol. 11 (3), pp. 130–149.

Philo, C 1995, 'Animals, geography and the city: Notes on inclusions and exclusions,' *Environment and Planning D: Society and Space*, vol. 3 (6), pp. 655–681.

Philo, C and Wilbert, C 2000, 'Animal spaces, beastly places: An introduction,' in C Philo and C Wilbert (eds), *Animal spaces, beastly places: New geographies of human-animal relations*, Routledge, London and New York.

Plumwood, V 2002, 'Prey to a crocodile,' *The Aisling Magazine*, vol. 30. Viewed September 16, 2014. www.aislingmagazine.com/aislingmagazine/articles/TAM30/Val-Plumwood.html.

Reilly, E 2009, 'Beach bullies: Coyotes on the prowl,' *The Hamilton Spectator*, February 26, p. A1.

Smith, M 2001, *An ethics of place: Radical ecology, postmodernity, and social theory*, State University of New York Press, Albany.

Spears, J 2009, 'Coyote that killed Beach dog marked for death,' *Toronto Star*, March 28, p. GT4. Viewed September 16, 2014. www.thestar.com/news/gta/2009/03/28/coyote_that_killed_beach_dog_marked_for_death.html.

Timm, R and Baker, R 2007, 'A history of urban coyote problems,' *Wildlife Damage Management Conference Proceedings*. Paper 76. Viewed September 16, 2014. http://digitalcommons.unl.edu/cgi/viewcontent.cgi?article=1075&context=icwdm_wdm confproc.

Whatmore, S 2002, *Hybrid geographies*, Sage, London.

White, L and Gehrt, S 2009, 'Coyote attacks on humans in the United States and Canada,' *Human Dimensions of Wildlife*, vol. 14 (6), pp. 419–432.

Wolch, J 1998, 'Zoöpolis,' in J Wolch and J Emel (eds), *Animal geographies: Place, politics and identity in the nature-culture borderlands*, Verso, London.

Wolch, J 2002, 'Anima urbis,' *Progress in Human Geography*, vol. 26 (2), pp. 721–742.

Wolch, J and Emel, J 1995, 'Bringing the animals back in,' *Environment and Planning D: Society and Space*, vol. 13 (6), pp. 632–636.

Wolch, J, West, K and Gaines, T 1995, 'Trans-species urban theory,' *Environment and Planning D: Society and Space*, vol. 13 (6), pp. 735–760.

Worcester, R and Boelens, R 2007, 'The co-existing with coyotes program in Vancouver BC,' *Proceedings of the 12th Wildlife Damage Management Conference*. Viewed September 16, 2014. http://digitalcommons.unl.edu/cgi/viewcontent.cgi?article=1078&context=icwdm_wdmconfproc.

10 Livelier livelihoods

Animal and human collaboration on the farm

Jody Emel,[1] *Connie L. Johnston, and*
Elisabeth (Lisa) Stoddard

In recent years, especially following Derrida (2008), Haraway (2008), and Braidotti (2013), a shift toward a more posthuman, postanthropocentric relational theorization of human and nonhuman animals has moved critical scholars and activists to tentatively rethink veganism as the only just position with respect to domesticated farmed animals (Pedersen 2011). These theorists are less focused on the similarities between nonhuman and human capacities, and are more interested in the ways in which we are mutually dependent, interrelated, and experience a shared vulnerability to socio-ecological threats (e.g., globalization and climate change) (Wolfe 2010; Head and Gibson 2012). We animals are more mutually dependent and less hierarchically positioned than Western cultural heritage has led us to believe. Materialist feminists inspire this reconsideration as well, because of their insistence on the inescapability of the ecological human body and its embeddedness within complex economic and biological potentialities and constraints (Alaimo and Hekman 2008).

If we recognize a flatter hierarchy, mutual dependency, and the limits of biology and landscape, can we North American, settler colonists, human bodies and beings, and critics of industrial animal production, reshape our entanglements with domestic farmed animals such that our relationships are more cooperative and respectful? We would like to explore the potential of what we call 'livelier livelihoods' for human and nonhuman animals in agriculture in order to consider the possibilities in between industrial animal agriculture and absolute veganism. In a livelier livelihood, species difference is valued and animal agency is enabled to benefit the lives and livelihoods of the humans and nonhumans on the farm, as well as to account for the quality of the surrounding ecosystems that we all depend upon.

If agriculture is based on human–nonhuman relationships, in which difference is valued within and between species and the environment encourages complex decision-making in both work and play, might we develop capacities to live lives with other animals that are more mutually beneficial, more just, and more respectful and conscious of each other's wants and needs? Immediately, all manner of questions arise regarding exploitation, power imbalances, and instrumentalities. But as Haraway argues regarding the latter, "[t]o be in a relation of use to each other is not the definition of unfreedom and violation."

And following, she writes that "[s]uch relations are almost never symmetrical ('equal' or calculable)" (2008: 74). The relationships of farmer and farmed animal; captor and captive; killer and one who is killed; or profiteer and commodity are clearly unequal. The equality of a relationship, however, depends upon more than a one-dimensional comparison. What is provided to the captive? Freedom to live one's life within boundaries established by others? Access to food and shelter, and opportunities for socializing and flourishing? What happens when the commodity is also a friend?

Within the multispecies "contact zone" as Haraway (2008) calls it, many interactions are possible. It depends upon the farmed animal and the farmer. Yet it also depends upon economy – whether it is a capitalist economy or an anarchical one, there is still a 'bottom line' based upon weather, feed costs, and multiple other variables. Can we practice a respectful, more just form of farming that recognizes the dynamically evolving and always relational preferences of farmed animals? Can we rethink farmed animal relations given the existing socioeconomic and ecological conditions within which we mutually survive? After all, we cannot go back in time to a pre-domestication period, nor can we ignore the material limits afforded by biogeochemical cycles and bodily processes. Context is critical to justice.

We contend that human–nonhuman animal relations on the farm can be more equal, more respectful, and much more lively and considerate of nonhuman animal subjectivity and intentional agency. Is this merely another attempt at animal welfare rather than a critical animal geography? We are still debating this amongst ourselves, but we encourage a critical approach to animal geographies that explores and debates the eating, use, and killing of other animals rather than relegating farmed animals to heritage museums and sanctuaries.

This essay is thus a thought experiment based on Stoddard's immediate experience with two kinds of pig farming: confined animal feeding operations (CAFOs) and permaculture-based operations in North Carolina. We argue that the CAFO model of pig production generally works to *overcome* animal agency, whereas the permaculture-based model generally works to *enable* animal agency (conscious intention).[2] The shared relationship between farmer and pigs in permaculture permits not only a "shared suffering" (Haraway 2008) or vulnerability, but also a 'livelier livelihood' in which animals can be understood as cooperators or collaborators. In this livelier livelihood, differences among and between species are valued, and nonhumans enable and are enabled by their human and nonhuman counterparts in socio-ecological systems. Pigs may be viewed as coworkers who interact with humans and other species to produce viable and potentially enriching lives for all. However, can pigs' lives be viable or lively if their reproduction is controlled and if they are eventually slaughtered for food? Once again, we are still debating this amongst ourselves, and, therefore, we explore this question throughout the remainder of this chapter.

Lively, laboring, relating animals

Haraway writes of how "working animals, including food- and fiber-producing critters, haunt [her] throughout" her book on "species meeting" (2008: viii). We too are haunted by these food- and fiber-producing animals and, after years of going round and round over ethics and rights, would like to explore the idea of a non-vegan alternative. All three of us have practiced veganism, sometimes for decades. Stoddard abandoned her veganism after ten or more years, following her work and experiences on various farms. She eats any meal others are kind enough to provide for her, and she buys non-CAFO-produced meat as a special treat and to support humane, sustainable farmers, perhaps once a month. Johnston calls herself a non-doctrinaire vegan and tries to pay attention to multiple variables of consumer ethics. Emel, after years of near and absolute veganism, is now a nondairy, non-CAFO eater who occasionally eats grass-fed beef and who, a few times a year, buys organic chicken for her family in order to support that alternative in her mainstream grocery. All three of us wonder: is it possible that we could coexist with many fewer, but more fulfilled, farmed animals that have lengthier and higher quality lives? Human and nonhuman animals have co-evolved over millennia, and it may not be possible for humans in some eco-systems to survive without consuming nonhumans and vice versa, problematic and full of ethical contradictions as the situation may be.

Our initial inspiration arose from the recognition that nearly all of us humans work to live. Could we consider food and fiber animals doing so? Rather than consigning them to farm sanctuaries (or even nonexistence), might we think of them as laboring for their livelihoods along with those of us who farm, eat, and wear their furs, skins, and wool? We conceive of the farmer and the animals as coworkers on the farm, similarly to Porcher and Schmitt (2012) in their inquiry into the work of 'dairy cows.' Cows invested their intelligence in the activities of work, they collaborated and cooperated, they showed a capacity to adapt to the constraints of work, and to use strategies to refuse such cooperation and even to cheat (ibid.). In the thought experiment attempted in this chapter, we assume both human and nonhuman animals require livelihoods and must work for them on the farm.

A 'livelihood' is a means of securing the basic necessities of life – food, water, shelter, and clothing (if necessary). Chambers and Conroy write that:

> a livelihood comprises people, their capabilities and their means of living, including food, income and assets. Tangible assets are resources and stores, and intangible assets are claims and access. A livelihood is environmentally sustainable when it maintains or enhances the local and global assets on which livelihoods depend, and has net beneficial effects on other livelihoods. A livelihood is socially sustainable which can cope with and recover from stress and shocks and provide for future generations.
>
> (1991: i)

As beings working for their shared livelihoods, all animals would receive occupational and safety protections, welfare nets in times of shortage (whether due to climate or economy), and other benefits. The farmer humans would provide food, shelter, medical and health care, protection from predators and violence, and other amenities to the nonhumans along the lines of the "ancient contract" described by Rollin (2008). In return, nonhumans would work as foragers, fertilizers, seed spreaders, mothers, caregivers, and food providers.

The use of the word 'contract' may not be apt here, as this is a legalistic concept that assumes two parties who are on relatively equal footing and capable of understanding the full extent of the commitment(s) being made. However, 'contract' does at least capture some sense of the bi- (or multi-)directional obligations. Additionally, the term 'food providers' is euphemistic, of course, and fails to capture the violent reality of what these animals give in return for being fed, sheltered, and protected from (other than human) predator violence. Some of them must die to become food and others must be surgically altered in order to prevent breeding, begging the question: is it possible for farmed animals to have good lives even if they are killed and eaten in the end? Even if they are allowed to live reasonably contented for several years before slaughter?

The comparative adjective 'livelier' suggests a relative improvement in the stimulations, energy, and freedoms associated with the farmer–nonhuman animal relationship. How can this be done? What keeps animals relatively busy, occupied, and content with their lives, and what is reasonable for farmers in terms of labor, capital, and markets? The task of knowing what animals want is difficult because there are so many species with vastly different wants and needs. In fact – the speed of life, the spaces traveled and inhabited, the various sensory faculties, the desires and practices – are so entirely different as to comprise multiple ontologies. Critical animal theorists Gruen and Weil remind us that "[f]or Derrida, the unknowability (beyond other and same) of [his] cat, as of any other subject, renders it impossible to quantify happiness or suffering. There can be no 'calculation' of the subject" (Gruen and Weil 2012: 480).

The pragmatics of the situation though require decisions about what happens on the farm – even the farm sanctuary. How do we, as humans who make many of the decisions about what happens on the farm (albeit as relational beings), better understand those other ontologies and their connected epistemologies? How do we communicate across worlds in order to allow and afford as much intentional agency and well-being as possible in our working collaborations? Let us look at current animal welfare/well-being sciences briefly before turning to the North Carolina pigs.

Farmed animal well-being, intentional agency, and work

What do animals want? How can we know? First, we have to acknowledge that this goal is a moving target. Our brains (human and nonhuman animal) change structure in response to stimulation and environment. There is no sharp distinction between inside and outside or nature and culture (Wilson 2004; Barad 2007). Recent work in cognitive science and ethology shows that the environment in which an animal lives affects the brain's ability to develop dendritic arbors, or neural networks (Gould and Gross 2002). Complex neural networks are critical for decision- and memory-making capabilities (ibid.). Animals who are engaged with their environments, who have to make daily decisions about social interactions, acquiring food, and choosing where to sleep, get cool, hunt, and play have more complex brains. These animals are capable of higher order decision-making than members of the same species who do not live in complex social and material environments (Gould and Gross 2002). This has implications for understanding animal well-being, animal agency in the context of place and space, and the foundation of human, animal, and ecosystem relations (Puppe *et al.* 2007). Thus, a livelier livelihood for pigs could only be understood as relational and dynamic.

Marian Stamp Dawkins, renowned animal behavior researcher, contends that understanding what makes animals healthy and "having what they want" requires carefully constructed welfare research in a commercial setting (2008). While we would agree that science with a capital 'S' might be more convincing to some, there is also a role for citizen farmer science and farmed animal sanctuary science. We believe that citizen science, via blogs, websites, farm fairs and conventions, and other approaches might complement, contradict, and complicate our understandings based on formal Science. Before discussing less formal approaches, we first briefly examine those that are more formally sanctioned.

Western scientific approaches to measuring welfare

For some time now, it has been widely acknowledged in the animal science community that welfare is a notoriously difficult thing to both define and measure – rather like human well-being (Miele and Bock 2007; Lay 2013). And some of those most respected and vocal scientists within the community argue that only measurable indicators of well-being should be considered because anything else is too subject to manipulation (Dawkins 2008), although many if not most do acknowledge that almost any measurement parameter involves some level of manipulation. Within this already muddied overall concept, how is animal agency being studied and/or measured in farmed animals? First of all, it should be noted that 'agency' (denoting conscious intention) is not a term that is widely used in animal welfare science. Terms such as 'preference' and 'motivation' are more commonly used to indicate some level of intention on an animal's part, although these terms in and of themselves do not signify it (see e.g., Bracke 2011). Below we discuss one of the few farmed animal welfare science publications that references agency, but first it will be helpful to situate this concept within the wider field.

Three basic (and frequently overlapping) approaches to farmed animal welfare exist in Western science. These three approaches base definitions and assessments on animal health or physiology, their environments, and their feelings. The physiologic approach has predominated in the U.S., but in recent years the feelings approach (seen as providing "some indication of how positive or negative an animal feels" (Duncan 2006: 14)), has increasingly gained acceptance. Among animal welfare scientists, many if not most consider welfare to be a suite of factors that interact in complex ways (Lay 2013).

One can see elements of the three approaches described above in current and recent studies of pig welfare and the ways that it may be measured. A major physiological welfare issue for pigs, one which garners much attention in farm animal science, is lameness and the associated pain (see e.g., Minton 2010; Johnson *et al.* 2011). Along with pain, stress is a major pig welfare issue in modern industrial animal agriculture, reflected in numerous publications (see e.g., James *et al.* 2013; Rault *et al.* 2013), and also in the current agendas of major U.S. farmed animal science research entities (Minton 2010; USDA 2013).

Beyond physiological welfare and for pigs in particular, there have been a number of recent scientific studies on responses to humans, motivation to engage in certain types of social and exploratory behavior, and preferences for such things as food, social contact, and different environmental components (e.g., straw, bedding, etc.). Overall, many of these studies find that it is difficult to make overarching statements about *types* of farms or pigs' preferences (Elmore *et al.* 2012; Temple *et al.* 2011). With regard to environmental enrichment, for example, Elmore *et al.* (ibid.) found that sows' motivation to access compost or straw depended on the method, quantity, and/or timing of provision. What all of these studies indicate is that there is likely considerable variation, not only with respect to individual animals, but also within individual farm environments and practices.

Welfare and agency

Although few animal welfare science publications reference agency, Špinka and Wemelsfelder's 'Environmental Challenge and Animal Agency' (2011) explicitly engages with agency and connects it to welfare. The authors pair the concept of agency with that of "competence," asserting that the two reflect "complementary aspects of the animal's engagement with novel challenges" (ibid.: 27). They go on to define competence as "the whole array of cognitive and behavioural experience, tools and strategies that an animal possesses at any given moment to deal with novel challenges," and agency as "the intrinsic tendency of animals to behave actively beyond the degree dictated by momentary needs, and to widen their range of competencies" (ibid.: 28).

Špinka and Wemelsfelder connect agency/competence to challenges that many animals would encounter in environments in which humans do not provide for them. They argue that, because of the challenges faced in terms of finding food, mates, and social stability in complex environments, animals must develop numerous types of learning skills (e.g., associative, instrumental, social, etc.), as well as the ability to deal with novel situations. (This ability has obvious application to climate and weather vulnerabilities that will be discussed below with regard to pigs on North Carolina farms.) Very generally, these skills and abilities are aspects of agency and competence and, to illustrate their functioning, the authors focus on three prominent ones: problem-solving, exploration, and play.

Citing a variety of supporting studies, Špinka and Wemelsfelder assert that competence-building activities and opportunities for exercising agency provide holistic benefits, stating that animals in farming environments also must cope, learn, and negotiate novel situations.

The relationship between agency and welfare has three dimensions. First, agency is inherently rewarding for animals, irrespective of any immediate, functional benefit. Second, animals receive positive reinforcement when they engage in competence-building activities. The exercise of agency in choosing to engage in these activities could further enhance the positive outcome, encouraging animals to seek out these types of activities. The final dimension incorporates the more obviously physiological, as the authors contend that a more psychologically satisfied animal will also likely be a healthier and less fearful animal, thereby reducing stress. Additionally, Špinka and Wemelsfelder illustrate how agency expression varies *between* individuals and also *within* an individual based on age, circumstance, etc. This is echoed by the studies on preference and motivation referenced above.

The foregoing review, although limited, indicates the myriad factors involved in farmed animal welfare and the ongoing challenges acknowledged by scientists to understanding animals' experiences.

Welfare, agency, and the wider society

A number of social scientists have also been interrogating the ways in which we humans may know other animals, broadly theorizing this topic (see e.g., Johnston 2008), but also using ethnography to examine the mutuality of relationships between humans and their animals (Lorimer 2006; Istomin and Dwyer 2010), examining animals' engagement as subjects with modern farming technologies (Holloway 2007), and participating in field studies of urban wildlife (Hinchliffe *et al.* 2005). Perhaps the most productive aspect of all of these research developments at this point is not that methodological breakthroughs have been achieved, but that new epistemological ground is being tested. Corresponding to the opinion of Špinka and Wemelsfelder noted above, we humans are (and probably will always be) limited in our ability to understand other animals. It is heartening that social, and not just natural, scientists are investigating ways of knowing nonhumans as livelier livelihoods are a collaboration based on social, economic, and biological relationships.

Citizen science is another source of information for the understanding of farm animal well-being. A variety of popular media outlets exist today that are venues for discussion of more 'animal-friendly' ways of farming. Regional organic farming organizations, including the Northeast Organic Farming Association (NOFA), hold conferences that generally include animal farmers who are experimenting with alternative systems that provide more freedom and stimuli for their animals. In 2009, NOFA hosted Joel Salatin, who conducted an all-day workshop on his multispecies grazing system approach. Blogs such as Motherearthnews, Sustainabletable, Backyardchickens, Localharvest, Thebeginningfarmer, and others offer a wealth of information for farmers with various operations who are testing an unconventional approach. The predilections of different breeds are discussed, as are housing types and relationships.

These blogs and other sources of alternative animal-raising information recognize that generalizations can only go so far. This is echoed, perhaps surprisingly, in much of the formal animal science reviewed above, suggesting that humans' close relationships with animals can be a key to maximizing the potential for agency and better welfare on particular farm environments. If farmers know their animals as individuals, then they can hopefully better perceive and attend to fluctuations in their conditions. However, as we have discussed, there are limits to knowing as well. We can never understand how it is to be a pig or know how their experiences *feel to them*. We can try, as fellow animals, to understand. We can hold ourselves accountable to them in changing, exploratory, respectful, and collectively oriented relationships – relationships rife with contradictions, vulnerabilities, violence, and uncertainties.

Livelier livelihoods on the farm

North Carolina is the second largest (after Iowa) hog farming state in the country, with nine to ten million pigs raised in CAFO farms each year (Nicole 2013). Over the last decade, there has also been a growth in small-scale alternative pork production, including permaculture-based farms (Curtis *et al.* 2011). We draw on six months of Stoddard's interviews with farmers and participant observation with humans and nonhumans on farms in North Carolina to examine the ways in which these two different types of farming systems enable and disable animal agency. We do so to consider how augmented animal agency may create the potential for livelier livelihoods. We also explore whether livelier livelihoods can benefit the interconnected and interdependent lives and well-being of farmed animals, farmers, and social and ecological systems in farm communities – especially in a changing climate and liberalizing global economy.

Enabling species-specific agency and behavior is one dimension of farmed animal well-being (Špinka and Wemelsfelder 2011). While pigs (like all animals) have unique and individual wants, needs, and desires, they also have a number of behaviors that they need to exercise as a species to ensure a high quality of life. These behaviors include exploring their environment through chewing and rooting (digging with their snouts) (Studnitz *et al.* 2007), building beds and nests (Gonyou 2001), wallowing in mud and shallow water (Bracke 2011), and interacting freely in social groups (e.g., playing, avoiding conflict) (Gonyou 2001).

Comparing these two farm types, we argue that permaculture farms tend to enable animal agency, while CAFO farms often function by disabling animal agency – with consequences for the farms' entire socio-ecological network. The CAFO and permaculture systems that we are comparing are not opposites, however. Instead, they fall along a range of uneven power relationships among humans and farmed animals. While there are significant differences in these two production systems, there are also a number of similarities. We examine the ways in which death, breeding, and movement are tightly controlled in both systems. We wonder if life can truly be lively for farmed animals if these controls are still in place. In addition, while we and others (Špinka and Wemelsfelder 2011) argue that augmented animal agency can benefit animal welfare, enabling animal agency alone does not ensure animal welfare (Lay 2013). We examine instances when enabling agency alone cannot help to promote the wellbeing of farmed animals or the socio-ecological systems of which they are a part. Here, we ask if increased regulations to protect nonhumans and humans can help, and if they are sufficient.

Permaculture with pigs

Permaculture is an integrated system of agro-ecological design based on enabling the wants and needs of nonhuman species in order to produce and procure food, water, fiber, energy, and profits for humans and nonhumans on the farm (Holmgren 2002). Pigs in permaculture systems are typically raised in groups of 5–50, in stocking densities of 3,000–14,000 square feet per pig, and are rotated throughout the farms' pasture, dirt, and woodlands (Holzer 2011).

While there is a significant diversity of practices within permaculture systems (Holzer 2011), there are a number of common philosophies and goals (Holmgren 2002). One common practice is conceptualizing and interacting with animals as active agents and coworkers who are critical to the function of the farm's ecological and social systems (Holzer 2011). Permaculturists conceptualize farmed animals as what Haraway calls "working subjects, not just worked objects" (2008: 80). Within this ideological framework, species difference and agency is valued, enabled, and animals are collaborated with as co-creators of the farm socio-ecological network. However, their lives and behaviors are only enabled as long as they remain valuable to the permaculture system (similar in some ways to human workers and their employers). When their behaviors are no longer beneficial to the farmer, pigs may be sold to other farms or sent to processing plants for slaughter (Jessie,[3] alternative swine farmer). We discuss slaughter in the context of livelier livelihoods at the end of this section.

While their work remains advantageous, pigs are encouraged to engage in behaviors that they love (rooting, digging, exploring), which can increase soil health, get rid of nonhuman animals and plants considered to be pests, improve biodiversity, sculpt the landscape, and more (Holzer 2011; Whitefield 2004). For example, farmers direct pigs to areas on the landscape where they want small ponds dug to hold water during rain events, to store water during droughts, and to support commercial crops like mushrooms, asparagus, and berries (Holzer 2011; Jessie, alternative swine farmer). The pigs dig and root in these areas to create wallows to regulate their body temperature, to protect themselves from insects and sunburn, and for enjoyment (Bracke 2011; Pat, alternative swine farmer) (see Figure 10.1). While pigs sculpt their desired baths, they are simultaneously contouring the land to hold water for use and storage – benefitting all

Figure 10.1 Pigs dig and root to create wallows (image © Elisabeth Stoddard).

of the species within the farm network who are dependent upon water (Holzer 2011; Jessie, alternative swine farmer).

Farmers also direct pigs to areas with invasive species or to fruit orchards to prevent the spread of mold from fallen fruit – putting the pigs to work, while satiating their omnivorous appetite and desire to explore (Holzer 2011; White-field 2004). Some farmers give pigs access to woodlands for farrowing (giving birth), satisfying pigs' desire to create nests (Gonyou 2001). Pregnant sows use branches, leaves, bark, and other materials to create nests (Mayer *et al.* 2002), while improving the health of profitable hardwoods by removing softwoods and plant life beneath the forest canopy (Talbott *et al.* 2004). The pigs also improve the soil health and fertility by foraging for tree nuts and running and playing. They mix manure and leaf litter, increasing the amount of organic matter in the soil, which benefits plant growth and those dependent upon plants for food and profit (Holzer 2011; Aeron, alternative swine farmer).

This production system, based on human–nonhuman collaborations, not only benefits the daily lives of pigs, farmers, and others in the network, it also increases the ability of humans and nonhumans on the farm to be resilient in an increasingly unpredictable climate. North Carolina experiences coastal storms, floods, and droughts (Robinson 2005), and the state may experience these events with an increasing intensity as our climate continues to change (Paerl *et al.* 2006;

Sallenger *et al.* 2012; Strzepek *et al.* 2010). The miniature ponds dug by pigs are critical for everyone on the farm, storing water in drought conditions and capturing excess water during floods (Holzer 2011; Jessie, alternative swine farmer). The soil health, developed by pigs' rooting, digging, and playing, increases the soil's ability to hold water and recover from drought more quickly (Holzer 2011; Bot and Benites 2005).

Multispecies partnerships on the farm can also enhance market resilience in a rapidly changing global economy. Farmers can direct pigs to contour the landscape to create various microclimates (e.g., climate surrounding a mini-pond or under a tree canopy) (Holzer 2011; Jessie, alternative swine farmer). Having multiple microclimates increases the range of species that the land can support and, therefore, the diversity of products the farm can sell (Holzer 2011; Whitefield 2004). This product diversity provides security for farmers as global markets and prices change (Holzer 2011). Microclimates improve biodiversity with multiple benefits for species on and off the farm, and they provide pigs with comfort in a range of temperatures, and with a diversity of food and landscapes to explore (Holzer 2011; Jessie, alternative swine farmer).

Creating livelier livelihoods on the farm through enabling animal agency can have significant benefits for the lives and well-being of the humans and nonhumans that make up the farm network. However, most management decisions on the farm remain primarily under the control of the farmers, such as where pigs move or which and how many pigs will live in social groups together. This uneven power relationship can have negative consequences for the lives and well-being of pigs. For example, pigs in permaculture systems typically reproduce through mating with one another and have more control over their breeding than pigs who are artificially inseminated in CAFOs. However, permaculture pigs' social and mating interactions are limited by the farmers' construction of their social groups and the fences that prevent them from moving freely to access or escape particular social interactions.

For example, one permaculture participant, Renee, explained her horror at finding one of her older sows injured, bleeding, and dehydrated after several young boars copulated with the sow repeatedly and unwantedly. The fencing around the acre of pasture and forest prevented the sow from escaping, and the boars prevented the sow from fleeing into the huts within the fenced area. The farmer had placed the older, less fertile sows and young boars together so that the young boars could learn about foraging and predators from the more experienced sows and to manage breeding to prevent overpopulation. Prior to this experience, this group of pigs lived together quite harmoniously. The farmer blamed herself for not restructuring the group as spring was approaching, knowing that the boars would soon be increasingly interested in mating.

We have argued that enabling animal agency can benefit nonhumans and humans on the farm. However, enabling animal agency alone does not ensure animal well-being (Lay 2013). For example, permaculture farms that face financial constraints will often enable pigs to breed more freely (Renee and Pat, swine alternative farmers). This increases the number of pigs on the farm, often beyond its carrying capacity, with negative consequences for humans and nonhumans within the farm network. Large populations of pigs on pasture, beyond the ecosystem's carrying capacity, can lead to the loss of topsoil, leaving clay soils that cannot grow vegetation (Pietrosemoli *et al*. 2012; Pietrosemoli and Menius 2009). Consequently, pig overpopulation can expose and damage tree roots, causing the death of trees (Pietrosemoli and Menius 2009; Renee, swine alternative farmer). Without tree cover, pigs lack protection from hot temperatures, which can lead to heatstroke and death (Guthrie 2012). Without access to dried vegetation in cold weather, pigs lack sticks, leaves, and other materials to stay warm, which can impact their health and welfare (Pietrosemoli and Menius 2009; Renee, swine alternative farmer). Without vegetative buffers, pigs can be exposed to standing water during rainstorms, pig manure can pollute local waterways, and as a result, pigs, humans, and other animals can be at risk for insect-borne disease transmission and illness (Terrell and Perfetti 1991; McKean and Hoyer 2006; Pietrosemoli and Menius 2009).

Regardless of their various opportunities and protections, most pigs will be killed when farmers determine that the pigs' working life is over. For example, while pigs' rooting and digging can be productive, it can also be destructive (Pietrosemoli *et al*. 2012). Pigs in the U.S., both domestic and feral, have played significant roles in the extinction of native plant species due to digging and rooting, with grave consequences for ecosystems and the humans and nonhumans who depend on them (Cole *et al*. 2012). Therefore, when pigs' work becomes destructive or the number of pigs exceeds the carrying capacity of the farm ecosystem, most pigs are sent to slaughter (Holzer 2011; Jessie, alternative swine farmer). Small, independent farmers do not often have immediate access to corporate-owned slaughter facilities. This can require independent farmers to transport their animals hours away to slaughterhouses, which can be very stressful for the animals (Taylor 2008; Schwartzkopf-Genswein *et al*. 2012).

Farmers, scientists, and others have analyzed what makes death less stressful and less painful for farmed animals (Blockhuis *et al*. 2013; Renee, alternative swine farmer). However, how might we analyze the slaughter of farmed animals within a conceptual and practical framework of creating a flatter human–nonhuman hierarchy? In a flatter hierarchy, is it inherently wrong and unjust for human animals to kill nonhuman animals? The authors of this paper and the authors of many other texts (e.g., Midgley 1998; Haraway 2008; Adams 2010) have debated these questions and have come to multiple conclusions. For example, Haraway (2008) suggests that we should focus less on whether or not we should kill, and more on what makes life killable: the institutionalized logic that deems some lives as worth living/saving and other lives as killable and disposable. We do not have the space to give these questions the attention they deserve, but we do believe that the ethics and justice of to kill or not to kill farmed (and other) animals is not a simple yes or no answer, but one situated in social, political, cultural, ethical, economic, and ecological contexts.

Enabling farmed animal agency is only one of many conceptual reframings that would be required to improve lives and livelihoods on farms. We need to reframe farm policies and regulations that often make it impossible for non-industrial-scale farmers to compete in the market without compromising management practices like overstocking animals on pasture. Not only do we need to change existing farm policies that speak directly to human lives and livelihoods, but also create policies that speak to the lives and livelihoods of nonhumans on farms who are generally regulated as worked objects and not working subjects (Haraway 2008).

CAFO swine production

In CAFO systems, growing pigs are raised in groups as they move through weaning to finishing, and pregnant and nursing sows are kept in individual crates (Gillespie and Flanders 2010). CAFO pigs typically live in barns with 1,000–3,000 other pigs in stocking densities of about 8–14 square feet per pig (Buhr *et al.* 2005). In CAFO houses the climate is controlled, feed and water are automated, and pigs' waste falls through slatted floors and is concentrated in open-air liquid lagoons (Gillespie and Flanders 2010).

CAFO pigs live in barren environments. The only material they encounter regularly are the slatted floors that they stand on, the bars that surround their pens or crates, and other pigs (Lay Jr. *et al.* 2000). In this system, pigs' desires to chew and root, to build nests and beds, to wallow, and to interact freely in social groups are thwarted (Studnitz *et al.* 2007; Buchholtz *et al.* 2000).

In CAFO systems, species difference and animal agency and behavior are not valued or seen as an asset to the farm. Within this ideological framework, individual farmers may conceptualize animals as subjects (Sidney, CAFO swine farmer), but the corporate CAFO system in which they are situated treats non-human (and sometimes human) laborers as worked objects. In swine CAFOs, pig agency and species-specific behaviors are seen as problems to overcome. Without leaves, dirt, and brush to root in, pigs root and chew on other pigs' tails, ears, and bodies (Buchholtz *et al.* 2000). Pigs experience painful abscesses, sores, and stress as a result (Schrøder-Petersen and Simonsen 2001). Piglets, bored in a barren environment, nurse constantly (Buchholtz *et al.* 2000; Harper, CAFO swine farmer). The piglets' teeth can cause injuries to the sow, who will try to periodically move away from her piglets for relief (Lewis *et al.* 2005; Harper, CAFO swine farmer).

These problems are managed by *disabling* pig agency and by incorporating bodily mutilations to intercept some of the effects of agency or as an antidote to more destructive agentic behavior. Piglets teeth are clipped and their tails are docked when they are a few days old (Gillespie and Flanders 2010; Lewis *et al.* 2005). Pigs with docked tails can experience acute stress and develop painful growths of nerve tissue (AVMA 2013). Teeth clipping can cause pain, abscesses, and infection. To prevent the sows from moving away from their piglets, the sows are kept in farrowing crates – forcing them to nurse (Gonyou 2001; Harper, industrial swine farmer). In this system, pig agency and behavior is a labor cost to farmers and a liability to treat, instead of a valued and respected asset on the farm.

Disabling pig agency not only impacts animal well-being and costs incurred by farmers, it has implications for the ability of industrial pigs and farmers to be resilient in extreme weather events. In 1999, Hurricane Floyd struck North Carolina, flooding hundreds of hog houses across the state's eastern coast (Barnes 2013). Unable to escape their crates and pens in order to move to high ground, 30,000 pigs drowned, along with three million chickens and turkeys also living in confinement (ibid.). In 2008, pigs living in CAFO systems in Iowa had a similar experience. Those who were confined drowned, while others released by farmers swam to higher ground (Verchick and Hall 2011).

Relatedly, while permaculture pigs are bred for robustness, industrial animals are bred for high production performance, increasing pigs' rates of growth and reproduction (Gillespie and Flanders 2010). High performance breeding can make animals more sensitive to extreme weather and variable temperatures, which can also limit their agency during disasters. Backlund *et al.* explain:

> [a]s performance levels (e.g., rate of gain, milk production per day, eggs/day) increase, the vulnerability of the animal increases and, when coupled with an adverse environment, the animal is at greater risk.... At very high performance levels, any environment other than near-optimal may increase animal vulnerability and risk.
>
> (2008: 64)

Limiting agency due to breeding or confinement not only reduces the resilience of animals in extreme weather, it can reduce the resilience of communities and ecosystems as well. The disposal of hundreds of thousands of farm animal carcasses poses serious threats to the health of humans, wildlife, and ecosystems (Ellis 2001; Miller 2012). The significant loss of animals also threatens farmers' livelihoods. Instead of working *with* the pigs to generate a farm with multiple landscapes and species to eat and for market, CAFO pigs' ability to work and act is removed, and they become the sole product. Their sole function is to reproduce or gain weight. This specialization can put farmers at risk with respect to unforeseeable changes in the climate and global economy (Holzer 2011; Strange 2008).

Finally, while animal agency creates value on permaculture farms, disabling animal agency creates additional costs for farmers and communities. In permaculture systems, pigs play and dig, turning manure into fertile soil. However, in industrial systems, the waste of sedentary pigs falls through the slatted floors and into open-air waste lagoons (Nicole 2013). This waste has become a liability for farmers and a health hazard for communities, with farmers facing lawsuits over the management of their lagoons (Neff 2013). Communities experience multiple threats to their health and quality of life from air, soil, and water pollution (Nicole 2013).

Globally, we are seeing a shift towards CAFO farms (Steinfeld *et al.* 2006). As we have demonstrated, nonhuman (and often human) agency is disabled in these systems, with serious consequences for the lives and livelihoods of all actants that comprise farm socio-ecological networks. As critical animal geographers, should we argue for policies that enable livelier livelihoods with occupational protections for human and nonhuman laborers? Or are we not questioning and challenging existing power relations critically enough?

Conclusion

We have put forth an argument for 'livelier livelihoods,' seeking to acknowledge the entangled lives and livelihoods of human and nonhuman animals that must sustain bodily existences. We have also presented information about the complexity of concepts such as farmed animal well-being and agency, as these concepts have myriad components and variability based on individual animals, their life stages, their human caretakers, and their environments. The relationship between us is complex as all living beings must, at some level, 'work' to sustain themselves and, at least in the present day, for many humans that involves raising, killing, and consuming nonhumans. This statement is not meant to be prescriptive, normative, or predictive, nor is it intended to naturalize those relationships. It is simply an observation of the way much of food production *is* structured in the present moment. Despite farmers' best intentions and high levels of agency on permaculture farms, the conditions for vulnerability, brutality, and violence still exist. As Haraway argues:

> [i]n eating we are most inside the differential relationalities that make us who and what we are.... There is no way to eat and not to kill, no way to eat and not to become with other mortal beings to whom we are accountable, no way to pretend innocence and transcendence or a final peace.
>
> (Haraway 2008: 295)

Nevertheless, we have argued that farming environments and practices that work *with* animals' known behavioral tendencies provide a way to approach more of a partnership type of on-farm relationship, and have contrasted these environments with farming systems that limit and oppose animals' behaviors.

It is clear that many improvements can be made in the lives of farmed animals and permaculture holds many benefits over industrial processes. Industrial systems cannot be modified enough to actually enable animal agency to the extent that animals experience lives worth living. The scale of industrial operations requires biosecurity and architectures of waste management that cannot be changed enough to actually produce a viable ecosystemic result. Recognizing the farmer as collaborator with the animals and as manager of an agro-ecological system that provides a number of public goods and services requires a paradigm shift rather than a tinkering with the industrial system. We recognize, however, that two primary and related limitations of *any* type of system are that they disrupt the animals' life-cycles and, ultimately, require them to experience an early death at the hands of humans and/or machines. Animals would not likely choose to be slaughtered, especially at or before the prime of their lives, and therefore this basic expression of agency will always be limited in a production setting. Additionally, and with respect to less extreme consequences, allowing unfettered animal agency would involve their choosing mates, raising offspring, forming their own social groups, etc. It is worth pointing out that animal domestication – farmed, companion, or otherwise – also fundamentally disrupts these life processes. Therefore, even in sanctuaries or loving homes, animals' choices are limited. Ultimately, it seems that this overarching relationship of domestication (often a complex entanglement of friendship and domination) is one we need to continue to interrogate.

Notes

1 This paper is the result of a collaborative effort. Authors are listed in alphabetical order.
2 In this chapter, we draw on animal science literature (Špinka and Wemelsfelder 2011) to define agency as conscious intention. The subject of animal and other nonhuman agency has been a subject of great debate within geography and the broader academy. It has been discussed in great detail elsewhere (e.g., Emel *et al.* 2002 and Buller 2014), and therefore we will not discuss it here.
3 The names of all participants, both the alternative and CAFO swine farmers, have been changed to protect participant identities.

References

Adams, C 2010, *The sexual politics of meat*, The Continuum International Publishing Group Inc, New York.

Alaimo, S and Hekman, S (eds) 2008, *Material feminisms*, Indiana University Press, Bloomington, IN.

AVMA 2013, *Welfare implications of teeth clipping, tail docking and permanent identification of piglets*, American Veterinary Medical Association, May 29, 2013. Viewed September 16, 2014. www.avma.org/KB/Resources/LiteratureReviews/Pages/Welfare-implications-of-practices-performed-on-piglets.aspx.

Backlund, P, Janetos, A, and Schimel, D 2008, 'The effects of climate change on agriculture, land resources, water resources, and biodiversity in the United States,' in M Walsh (ed.), *Synthesis and Assessment product*, The National Science and Technology Council and the U.S. Climate Change Science Program, Washington, DC.

Barad, K 2007, *Meeting the universe halfway: Quantum physics and the entanglement of matter and meaning*, Duke University Press, Durham, NC.

Barnes, J 2013, *North Carolina's hurricane history*, 4th Edition, University of North Carolina Press, Chapel Hill.

Blokhuis, HJ, Miele, M, Veissier, I, and Jones, B 2013, *Improving farm animal welfare: Science and society working together: The welfare quality approach*, Wageningen Academic Publishers, Wageningen, the Netherlands.

Bot, A and Benites, J 2005, *The importance of soil organic matter: Key to drought-resistant soil and sustained food production*, Food and Agriculture Organization of the United Nations (FAO), Rome.

Bracke, MBM 2011, 'Review of wallowing in pigs: Description of the behaviour and its motivational basis,' *Applied Animal Behaviour Science*, vol. 132 (1) pp. 1–13.

Braidotti, R 2013, *The posthuman*, Polity Press, Cambridge.

Buchholtz, CB, Lambooij, B, Maisack, C, Marin, G, van Putten, G, Schmitz, S, and Teuchert-Noodt, G 2000, 'Ethological and neurophysiological criteria of suffering in special consideration of the domestic pig,' Paper read at Workshop of the International Society for Animal Husbandry, Bielefeld, Germany.

Buller, H 2014, 'Animal geographies I,' *Progress in Human Geography*, vol. 38 (2), pp. 308–318.

Buhr, B, Holtkamp, D, Brumm, M, and Kliebenstein, J 2005, *Economic analysis of pig space: Comparison of production system impacts*, National Pork Board and University of Minnesota, St. Paul, MN.

Chambers, R and Conroy, G 1991, 'Sustainable rural livelihoods: Practical Concepts for the 21st century,' IDS Discussion Paper 296.

Cole, RJ, Litton, CM, Koontz, MJ, and Loh, RK 2012, 'Vegetation recovery 16 years after feral pig removal from a wet Hawaiian forest,' *Biotropica*, vol. 44 (4), pp. 463–471.

Curtis, J, McKissick, C and Spann, K 2011, *Growth in the niche meat sector in North Carolina*, Center for Environmental Farming Systems, Goldsboro, NC.

Dawkins, MS 2008, 'What is good welfare and how can we achieve it?,' in MS Dawkins and R Bonney (eds), *The future of animal farming: Renewing the ancient contract*, pp. 73–82, Blackwell, Oxford.

Derrida, J 2008, *The animal that therefore I am*, Fordham University Press, New York.

Duncan, IJH 2006, 'The changing concept of animal sentience,' *Applied Animal Behaviour Science*, vol. 100, pp. 11–19.

Ellis, DB 2001, *Carcass disposal issues in recent disasters, accepted methods, and suggested plan to mitigate future events*, Southwest Texas State University, San Marcos.

Elmore, RP, Garner, JP, Johnson, AK, Kirkden, RD, Patterson-Kane, EG, and Pajor, EA 2012, 'Differing results for motivation tests and measures of resource use: The value of environmental enrichment to gestating sows housed in stalls,' *Applied Animal Behaviour Science*, vol. 141 (1–2), pp. 9–19.

Emel, J, Wilbert, C, and Wolch, J 2002, 'Animal geographies,' *Society and Animals*, vol. 10 (4), pp. 407–412.

Gillespie, JR and Flanders, FB 2010, *Modern livestock & poultry production*. 8th Edition, Cengage Learning Inc., Clifton Park, NY.

Gonyou, HW 2001, 'The social behavior of pigs,' in LJ Keeling and HW Gonyou (eds), *Social behavior in farm animals*, CABI Publishing, New York.

Gould, E and Gross, C 2002, 'Neurogenesis in adult mammals: Some progress and problems,' *The Journal of Neuroscience*, vol. 22 (3), pp. 619–623.

Gruen, L and Weil, K 2012, 'Introduction: Feminists encounteering animals,' *Hypatia*, vol. 27 (3), pp. 492–526.

Guthrie, T 2012, 'Heat stress in pigs,' Michigan State University Extension. Viewed September 16, 2014. http://msue.anr.msu.edu/news/heat_stress_in_pigs.

Haraway, D 2008, *When species meet*, University of Minnesota Press, Minneapolis.

Head, L and Gibson, C 2012, 'Becoming differently modern: Geographic contributions to a generative climate politics,' *Progress in Human Geography*, vol. 36 (6), pp. 699–714.

Hinchliffe, S, Kearnes, M, Degen, M and Whatmore, S 2005, 'Urban wild things: A cosmopolitical experiment,' *Environment and Planning D: Society and Space*, vol. 23, pp. 643–658.

Holloway, L 2007, 'Subjecting cows to robots: Farming technologies and the making of animal subjects,' *Environment and Planning D: Society and Space*, vol. 25 (6), pp. 1041–1060.

Holmgren, D 2002, *Principles & pathways beyond sustainability*, Holmgren Design Services, Hepburn, Vic.

Holzer, S 2011, *Sepp Holzer's permaculture: A practical guide to small-scale, integrative farming and gardening*, Chelsea Green Publishing, White River Junction, VT.

Istomin, KV and Dwyer, MJ 2010, 'Dynamic mutual adaptation: Human-animal interaction in reindeer herding pastoralism,' *Human Ecology*, vol. 38, pp. 613–623.

James, BW, Tokach, MD, Goodband, RD, Nelsson, JL, Dritz, SS, Owen, KQ, Woodworth, JC, and Sulabo, RC 2013, 'Effects of dietary L-carnitine and ractopamine HCl on the metabolic response to handling in finishing pigs,' *Journal of Animal Science*, vol. 91, pp. 4426–4439.

Johnston, C 2008, 'Beyond the clearing: Towards a dwelt animal geography,' *Progress in Human Geography*, vol. 32 (5), pp. 633–649.

Johnson, A, Coetzee, J, Stalder, L, Karriker, I, and Millman, S 2011, 'Pain: A sow lameness model,' *Journal of Animal Science*, vol. 89 (E2), p. 48.

Lay, Jr., D 2013, 'Safeguarding well-being of food producing animals,' Livestock Behavior Research Unit Publications. Viewed January 27, 2014. www.ars.usda.gov/research/publications/publications.htm?seq_no_115=300934.

Lay, Jr., DC, Haussmann, MF, and Daniels, MJ 2000, 'Hoop housing for feeder pigs offers a welfare-friendly environment compared to a nonbedded confinement system,' *Journal of Applied Animal Welfare Science*, vol. 3 (1), pp. 33–48.

Lewis, E, Boyle, LA, Brophy, P, O'Doherty, JV, and Lynch, PB 2005, 'The effect of two piglet teeth resection procedures on the welfare of sows in farrowing crates. Part 2,' *Applied Animal Behaviour Science*, vol. 90 (3), pp. 251–264.

Lorimer, H 2006, 'Herding memories of humans and animals,' *Environment and Planning D: Society and Space*, vol. 24, pp. 497–518.

Mayer, JJ, Martin, FD, and Brisbin, Jr., IL 2002, 'Characteristics of wild pig farrowing nests and beds in the upper Coastal Plain of South Carolina,' *Applied Animal Behaviour Science*, vol. 78 (1), pp. 1–17.

McKean, J and Hoyer, S 2006, 'Insect bites can mean lower income for pork producers,' Iowa State University Extension, Extension News, June 22. Viewed September 17, 2014. www.extension.iastate.edu/news/2006/jun/152201.htm.

Midgley, M 1998, *Animals and why they matter*, University of Georgia Press, Athens, GA.

Miele, M and Bock, B 2007, 'Competing discourses of farm animal welfare and agri-food restructuring,' *International Journal of Sociology of Food and Agriculture*, vol. 15 (3), pp. 1–7.

Miller, L 2012, 'Global lessons from FMD outbreaks: Implications for the US,' Wide Area Recovery and Resiliency Program, Denver, CO.

Minton, JE 2010, 'Statement of Issues and Justification,' *NC1029: Applied Animal Behavior and Welfare (NCR131)*. Viewed January 27, 2014. www.nimss.umd.edu/lgu_v2/homepages/home.cfm?trackID=13156.

Neff, J 2013, 'Hundreds file complaints over hog-farm waste,' *News and Observer*. Viewed September 17, 2014. www.newsobserver.com/2013/07/07/3015749/hundreds-file-suits-over-hog-farm.html.

Nicole, W 2013, 'CAFOs and environmental justice: The case of North Carolina,' *Environmental Health Perspectives*, vol. 121 (6), pp. 182–189.

Paerl, HW, Valdes, LM, Joyner, AM, Pieierls, BL, Piehler, MF, Riggs, SR, Christian, RR, Eby, LA, Crowder, LB, Raums, JS, Clesceri, EJ, Buzzelli, CP, and Luettich, Jr., R 2006, 'Ecological response to hurricane events in the Pamlico Sound System, North

Carolina, and implications for assessment and management in a regime of increased frequency,' *Estuaries and Coasts*, vol. 29 (6A), pp. 1033–1045.

Pedersen, H 2011, 'Release the moths: Critical animal studies and the posthumanist impulse,' *Culture, Theory and Critique*, vol. 52 (1), pp. 65–81.

Pietrosemoli, S and Menius, L 2009, *Importance of vegetative ground cover maintenance in pastured swine operations*, Center for Environmental Farming Systems, Greensboro, NC.

Pietrosemoli, S, Green, J, Bordeaux, C, Menius, L, and Curtis, J 2012, *Conservation practices in outdoor hog production systems: Findings and recommendations from the center for environmental farming systems*, Center for Environmental Farming Systems, Raleigh, NC.

Porcher, J and Schmitt, T 2012, 'Dairy cows: Workers in the shadows?,' *Society and Animals*, vol. 20, pp. 39–60.

Puppe, B, Ernst, K, Schön, PC, and Manteuffel, G 2007, 'Cognitive enrichment affects behavioural reactivity in domestic pigs,' *Applied Animal Behaviour Science*, vol. 105 (1), pp. 75–86.

Rault, J, Mack, LA, Carter, CS, Garner, JP, Marchant-Forde, JN, Richert, BT, and Lay, Jr., DC 2013, 'Prenatal stress puzzle, the oxytocin piece: Prenatal stress alters the behaviour and autonomic regulation in piglets, insights from oxytocin,' *Applied Animal Behaviour Science*, vol. 148 (1/2) pp. 99–107.

Robinson, PJ 2005, *North Carolina weather and climate*, The University of North Carolina Press, Chapel Hill, NC.

Rollin, B 2008, 'The ethics of agriculture: The end of true husbandry,' in MS Dawkins and R Bonney (eds), *The future of animal farming: Renewing the ancient contract*, pp. 7–20, Blackwell Publishing, Oxford.

Sallenger Jr., AH, Doran, KS, and Howd, PA 2012, 'Hotspot of accelerated sea-level rise on the Atlantic coast of North America,' *Nature Climate Change*, vol. 2 (12), pp. 884–888.

Schrøder-Petersen, DL and Simonsen, HB 2001, 'Tail biting in pigs,' *The Veterinary Journal*, vol. 162 (3), pp. 196–210.

Schwartzkopf-Genswein, KS, Faucitano, L, Dadgar, S, Shand, P, González, LA and Crowe, TG 2012, 'Road transport of cattle, swine and poultry in North America and its impact on animal welfare, carcass and meat quality: A review,' *Meat science*, vol. 92 (3), pp. 227–243.

Špinka, M and Wemelsfelder, F 2011, 'Environmental challenge and animal agency,' in MC Appleby, JA Mench, IAS Olsson, and BO Hughes (eds), *Animal welfare*, CAB International, Cambridge, MA.

Steinfeld, H, Gerber, P, Wassenaar, T, Castel, V, Rosales, M and De Haan, C 2006, *Livestock's long shadow*, Food and Agriculture Organization of the United Nations (FAO), Rome.

Strange, M 2008, *Family farming: A new economic vision*, 2nd Edition, University of Nebraska Press, Lincoln, NE.

Strzepek, K, Yohe, G, Neumann, J, and Boehlert, B 2010, 'Characterizing changes in drought risk for the United States from climate change,' *Environmental Research Letters*, vol. 5 (4).

Studnitz, M, Jensen, MB, and Pedersen, LJ 2007, 'Why do pigs root and in what will they root?: a review on the exploratory behaviour of pigs in relation to environmental enrichment,' *Applied Animal Behaviour Science*, vol. 107 (3), pp. 183–197.

Talbott, CW, Reddy, GB, Raczkowski, C, Barrios, T, Matlapudi, M, Coffee, A, and

Andrews, J 2004, 'Environmental impact of integrating crop and sylvan systems with swine,' *Journal of Animal Science*, vol. 82 (Supplement 1), p. 303.

Taylor, DA 2008, 'Does one size fit all?: Small farms and US meat regulations,' *Environmental health perspectives*, vol. 116 (12), pp. A529–A531.

Temple, D, Manteca, X, Velarde, A, and Dalmau, A 2011, 'Assessment of animal welfare through behavioural parameters in Iberian pigs in intensive and extensive conditions,' *Applied Animal Behaviour Science*, vol. 131 (1/2), pp. 29–39.

Terrell, CR and Perfetti, PB 1991, 'Animal wastes,' in SC Service (ed.), *Water quality indicators guide: Surface waters*, Diane Publishing, Darby, PA.

United States Department of Agriculture (USDA) 2013, 'Livestock behavior research mission'. Viewed September 22, 2014.

Verchick, RRM and Hall, A 2011, 'Adapting to climate change while planning for disaster: Footholds, rope lines, and the iowa floods,' *Brigham Yound University Law Review*, vol. 6.

Whitefield, P 2004, *Earth care manual: A permaculture handbook for Britain and other temperate countries*, Permanent Publications, East Meon, UK.

Wilson, EA 2004, *Psychosomatic: Feminism and the neurological body*, Duke University Press, Durham, NC.

Wolfe, C 2010, *What is Posthumanism?*, University of Minnesota Press, Minneapolis.

11 En-listing life

Red is the color of threatened species lists

Irus Braverman

Not all threatened species are created equally.

(Mike Hoffmann, IUCN, interview)

The Cebu flowerpecker *Dicaeum quadricolor* is a native bird of the island of Cebu in the Philippines (BirdLife International 2014a). The species was listed as Extinct in the first bird Red Data Book in the 1960s and was placed on the International Union for Conservation of Nature (IUCN) Red List of Threatened Species™ (hereafter, the Red List) in 1988. The bird's surprise rediscovery in 1992 resulted in the species' reassessment as Critically Endangered in 1994 (BirdLife International 2014b). Currently, the flowerpecker is believed to be one of the ten rarest birds in the world. Although the bird was observed in 2014, no one has seen one long enough to take a photograph, and not much other information has been recorded about the flowerpecker. There is, however, the long and convoluted history of this bird's listing.

Errors in species listings are not uncommon, Thomas (Tom) Brooks, head of IUCN's Science and Knowledge Unit, explains in an interview. In this case, he continues, the incorrect Extinct listing "turned into a self-fulfilling prophecy," effectively producing what has been referred to by scientists as a "Romeo Error" (Collar 1998). "In giving up on it," Mike Hoffmann – Senior Scientific Officer at the IUCN Species Survival Commission – explains, "the species actually really does go extinct. Because once there is no attention to it, all those threats actually do wipe it out" (interview). "The assessment in the 1960s was wrong," Hoffmann continues:

> If it had been conducted more carefully, it wouldn't have said that the species was extinct [when], obviously, it wasn't. The species would have then maintained attention and an effort towards safeguarding both the species and the biodiversity of Cebu over the past fifty years.
>
> (Interview)

As it is, however, the flowerpecker was most likely down-listed from the Extinct category only to be re-listed in this category in the not so distant future.

As the flowerpecker example shows, the act of listing threatened species impacts the life and death of actual animals. Indeed, beyond their descriptive and declarative functions, threatened species lists prescribe a series of material effects on very particular animal bodies; they also normalize and regulate conservation and related actions on the part of specific humans and networks. While recognizing these functions and effects on the individual level, this chapter focuses on the management of life at the level of the biological *species*, what Foucault refers to as biopolitics, as distinct from (yet entangled and coproduced with) anatomopolitics (Foucault 1990). In other words, I examine how the practices of assessing and listing nonhuman species translate into particular knowledges of species, as connected with, yet distinct from, knowledges of individual animals and populations.

My focus on species admittedly goes against the grain of animal geography's emphasis of late on individual animals rather than on "collectivities such as 'animals,' 'species,' and 'herds'" (Bear 2011: 297). The problem with such exclusive attention to the individual animal is that it disregards myriad non-human life worlds, networks, systems, and relationalities that do not necessarily flow directly from the individual scale. I am not saying that critical animal geographies should not take the individual lens seriously; I am merely cautioning against too widespread an uptake of Bear's recommendation, particularly at this sensitive time. Accordingly, this chapter offers a critical account of biopolitical projects that focus on the *species* as a unique entity of government, thereby revealing a range of power dynamics that would neither be operative nor apparent when discussing the individual animal and the population.

The project of governing species sits somewhere between that of governing individuals and that of governing statistical populations – and corresponds with both. Unlike Foucault's abstract population (which, I should point out, is different from the understanding of a population in the conservation context, typically as a unit smaller than a species), a species has a face and a context; it is situated – as becomes clear from the narratives of conservation experts. Put differently, thinking and governing through species regimes enables both an abstraction – a grid over the Linnaean kingdoms (Foucault 1970) – and an embodiment – a personification of ecosystems, habitats, and populations. Since humans understand themselves primarily as a species and therefore both relate to, and differentiate themselves from, other species – it is important to critically examine this lens and the work that it performs in the world.

For conservation biologists, the species is the foundational ontological unit for knowing and calculating life, or viability (Braverman 2014 and 2015; Sandler 2012). Biermann and Mansfield reflect on the perspective of conservation experts that:

Managing individual nonhuman lives is meaningless in responding to the crisis of biodiversity loss; individual lives acquire meaning only when they advance the long-term well being of the broader population or are essential to sustaining key biological processes, especially evolution.

(2014: 264)

According to this way of thinking, the death of an individual gains meaning according to the level of endangerment of the species: once on the brink of extinction, for example, the individual becomes larger than a singular life, and her or his death is therefore more than a singular death: it becomes the death of a life form, the death of nature (Braverman 2015).

At the same time, the deaths of so many other life forms who are not rare, charismatic, or visible enough to warrant the 'threatened' designation fall outside the range of protections established by the list, or outside the list altogether. Such life forms are effectively 'list-less': incalculable, unmemorable, and thus killable. Toward the end of this chapter, I argue that the conservation value of a species is defined through its inclusion and rank in an increasing number of lists and that the power of such lists is constantly eroded as new lists take their place in defining what is even more threatened, endangered, or extinct. Foucault (2003) refers to this project of differentiating between what must live and what must die as "racism." Although he restricts his analysis to humans and to the complex histories of the animalization of racialized others, I find that this framework is crucial to understanding the process of human exceptionalism and speciation and is thus highly relevant also in the context of critical animal geographies.

My project illuminates the immense regulatory power of lists and their heightened focus on, and differentiation of, life. Specifically, I argue that in addition to reinforcing the biopolitical differentiation between perceivably distinct nonhuman species, threatened species lists also reinforce the biopolitical differentiation between human and nonhuman species, with the human never being subject to the threatened list. Such a differentiated, or racial, treatment of the life and death of species through their en-listing, down- and up-listing, multi-listing, and un-listing translate into the positive protection and active management of nonhumans. Threatened species lists are thus biopolitical technologies in the battle against biological extinction. Listing threatened species becomes a way to affirm – and justify – that life which is more and most important to save (Braverman 2015).

While most of the biopolitical work in geography is centered on thanatopolitics or necropolitics, this project brings into focus an affirmative biopolitics (Rutherford and Rutherford 2013: 426), namely "the ways in which biopolitics can be more about life than death, about inclusion rather than exclusion" (2013: 429). What happens to those list-less lives that fall outside the realm of the threatened list does not figure within this account, which focuses instead on the viability of the listed. But such a focus on the affirmative does not entail a disavowal of death. Quite the contrary, as Biermann and Mansfield argue:

> to *make live* does not mean to avoid death altogether but to manage death at the level of the population. In a biopolitical regime, death is transformed into a rate of mortality, which is open to intervention and management. This transformation erases the fact that not all life is equally promoted.
>
> (2014: 259, italics in original)

For the list-less, the rule is typically the non-application of protection and the phasing out of support, although it can include much more explicitly sovereign methods when pertaining to certain species, especially those that threaten the purity of the listed (e.g., Gila and rainbow trout, or crested and marine toads; Braverman forthcoming). But while the color of the Red List is intended to alert to the dire state of those species that are listed as threatened and the intensified management of their mortality rates, it fails to alert to those species and individual animals who have been marginalized in the process of saving the chosen ones (Braverman 2015). Inspired by Tania Li (2009: 67), I conclude with a question: is it possible for social forces to mobilize in a wholly "make live" direction?

A few words on the structure of this chapter. After a brief discussion of biopolitics and its application to nonhumans, I examine the work that lists, and threatened lists in particular, purport to do, and their appeal from a regulatory standpoint. Next, I discuss the IUCN Red List and explore this list's unique category of Extinction. I conclude with a short discussion of interrelated incentives such as the lists of the Alliance for Zero Extinction (AZE) and EDGE of the Zoological Society of London. My heavily empirical and ethnographic focus in this chapter conveys the biopolitical project that lies at the heart of conservation biology, thereby contributing to critical animal geographies of nonhuman species.

Biopower in conservation biology

> All kinds of things become more interesting once we stop assuming that 'we' are the only place to begin and end our analysis.
>
> (Hinchliffe and Bingham 2008: 1541)

Michel Foucault's concept of "biopower" helps make sense of conservation biology's extensive use of species ontology, its fundamental trust in numbers, and its focus on calculations of rarity in practices of listing life. In the pre-modern period, sovereign power was characterized by the "the right to decide life and death," that is, the right to *take* life or *let* live (Foucault 1990: 135–136). Foucault argues that this ancient right has been replaced by a "power to *foster* life or *disallow* it to the point of death" (1990: 138, italics added). He defines this new "power over life" – which he sees as emerging in the eighteenth century with the development of bourgeois society and capitalism – as "biopower." In his words:

> Power would no longer be dealing simply with legal subjects over whom the ultimate dominion was death, but with living beings, and the mastery it would be able to exercise over them would have to be applied at the level of life itself; it is the taking charge of life, more than the threat of death, that gave power its access even to the body.
>
> (1990: 142–143)

Power, Foucault argues, no longer has death as its focus, but rather the administration of the living: "Such a power has to qualify, measure, appraise, and hierarchize, rather than display itself in its murderous splendor" (1990: 144).

Although Foucault uses the term 'biopower' to describe the project of governing *human* bodies, populations, and life (see also Rabinow and Rose 2006; Rose 2001), my work draws on a growing scholarship that expands this notion to the governing of nonhuman animal species and populations (Friese 2013; Haraway 2008; Rutherford and Rutherford 2013; Shukin 2009; Wolfe 2013). Within this scholarship, limited attention has been paid to the role of race in the biopolitical differentiation of nonhuman life (but see Biermann and Mansfield 2014: 261). Although the project of racism, as Foucault defines it is crucial for explaining the distinction between list-less and listed life, my application of it is different than Foucault's, as I shall explain shortly.

Death, in this context, is a means to foster life. In Foucault's words: "The enemies who have to be done away with are not adversaries in the political sense of the term; they are threats, either external or internal, to the population" (2003: 256). Foucault refers to the break between the livable and killable as "racism." According to this definition, the death of the other improves life as a whole:

> racism justifies the death-function in the economy of biopower by appealing to the principle that the death of others makes one biologically stronger insofar as one is a member of a race or population, insofar as one is an element in a unitary living plurality.
>
> (2003: 258)

Unlike in Foucault, however, in this case the "list-less" population is ostensibly that which is *not threatened*, and not necessarily that which threatens. Rather than posing a biopolitical threat to the flourishing of listed populations, list-less populations simply remain killable, whereas the threatened ones are elevated into a grievable status. So while the vulnerability of certain forms of nonhuman species life is what triggers human protection, the major threats that in fact create such vulnerabilities are depersonalized and abstracted. Terms such as 'deforestation,' 'climate change,' and 'habitat destruction' conceal the underlying assumption of conservationists that, for the most part, it is the conduct of *Homo sapiens* that is responsible for the other species' increased risk of extinction. Such an omnipresent threat of violence by humans looms over the threatened (note the passive voice) lists, but it is rarely made explicit. Simultaneously, certain list-less species are downgraded to the category of 'invasive,' 'hybrid,' or 'nuisance,' posing a second, this time more typically biopolitical, threat to the purity of the protected species. These interspecies threats become subject to forms of control by humans, such as elimination or purity management (Braverman 2015). To reiterate: threatened species lists are biopolitical technologies in their reinforcement of underlying species ontologies – and in their distinction between threatened nonhumans and never-threatened humans in particular; such lists are also about creating, calculating, and re-performing that line between nonhuman lives that are killable and those lives that should be cultivated.

Lists and threatened species lists

Let me now take one step back to consider the definition of a list and what, in particular, are the history and functions of threatened species lists. The *Oxford English Dictionary* defines a 'list' as "a number of connected items or names written or printed consecutively, typically one below the other" (2013). The word 'list' originates from 'border,' 'edge,' 'boundary' (from Old High German *lista*; *OED* 2013), but it also means lust and desire, or inclination. Dating back to Old English from before the twelfth century, *hlyst* also means 'to listen' (*OED* 2013).

A grocery list, kill lists, sex offender lists, and lists of threatened species – all share certain properties that define them as such: they are consecutive configurations of discrete items linked by a common goal that assigns them meaning and functionality. Lists name, classify, document, and simplify; they aspire to comprehensiveness, comparability, consistency, and uniformity, and are structured so as to delineate boundaries and produce authority and focused awareness. Making a list is thus a way to make something apparent (or heard, recall *hlyst*) that is not otherwise so. Related to and drawing upon all these functions, lists also standardize and regulate. Whereas all lists rely on various forms of classification, effectively "sorting things out" (Bowker and Star 1999) – some do more than that: they *prioritize*. With such lists, not only the listed items but also their particular order is significant.

Threatened species lists emerged in the 1960s and proliferated especially from the 1990s on. Today, conservationists routinely utilize lists of threatened species as powerful technologies in the battle against nonhuman extinction. Such lists share a few characteristics: they are typically a *scientific* method for highlighting those *species* under higher *extinction risk* with the explicit or implicit goal of focusing attention on *conservation measures* designed to protect them (Possingham *et al.* 2002: 503; italics added). This section will explore the properties of conservation lists that make them such ubiquitous tools of conservation and into such effective biopolitical technologies.

Mike Parr is the Chair of the Alliance for Zero Extinction as well as the Vice President of Planning and Program Development at the American Bird Conservancy. Parr ties the tendency to list threatened animals to our primordial function as hunters and gatherers. In his words: "We [wouldn't] want to kill the last one because we knew that if we did that, we wouldn't be able to eat" (Parr, interview). "If you're not actually going to hunt it," Parr adds, "a very nice surrogate is to make a list of it." The act of listing is thus not only a way of documenting life but also a way of knowing it – both in the sense of experiencing intimate physical contact with it and in being able to consume it. This explanation helps Parr argue for the importance of lists beyond their economic value. "There's a value to it that is not economic; it's intangible, probably," he tells me in our interview, concluding: "if we don't do something about it now, people will find that hole that's left in our collective soul and be mournful of it" (interview). From Parr's perspective then, acts of listing life are tied to our essential biophilic needs and desires as humans. In the process of upgrading the animal from huntable to savable, it is simultaneously elevated from the status of killable to that of the grievable.

John Lamoreux of the National Fish and Wildlife Foundation explains that "birders are famous for making lists: you have to be able to see what you saw. There's almost a listing mentality" (interview). But beyond its routine application, Lamoreux points out that the list is also important as a "rallying cry." "As an organization, you know what you stand for if you make a list of what's important," he tells me. "The whole reason you go into this is you want something added by doing this," he continues. "But you don't even start these exercises if you don't have some clear idea of what's missing, or what would get added value, or what would get raised attention, as needed" (interview). The list is thus a technology for differentiation.

Another IUCN scientist, Red List Unit Programme Officer Rebecca Miller, focuses on the functionality of lists. She writes: "The principle aim of a threatened species assessment is to estimate a species' risk of extinction in a comparable, repeatable, transparent, and objective manner" (Miller 2013: 191). According to Miller, threatened lists quantify the magnitude of the contemporary extinction crisis so as to monitor the status of biodiversity and measure current trends; they draw attention to the plight of threatened species and help mobilize political and public support for conservation measures; and they help guide conservation efforts into taking action where it is most needed (2013: 191). Finally, Miller argues that the capacities of threatened species lists to quantify, draw attention, and guide action are what have made them such powerful tools for mobilizing scientific, political, and public support (2013: 191).

The use of species as the foundational unit of threatened lists – effectively rendering them the "currency of conservation" (Lamoreux, interview) – is not only ideological but also pragmatic. First, species are the most common and easily measured unit for assessing the state of biodiversity. Moreover, threatened species are "among the most visible and easily understood symbols of the rising tide of extinctions," making them an "emotive and politically powerful measurement of biodiversity loss" (Miller 2013: 192; see also Wilson 1992; Wilcove 2010). In other words, species are the personalization – the individuation even – of populations and ecosystems. Using the species scale thus enables conservationists to put a face onto less apparent extinction processes and losses.

The lists' utilization of the species unit not only implies equality among species but also their comparability and homogeneity. The Red List, for example, is "applied to grasshoppers as well as blue whales," Lamoreux tells me. "There's something about the applicability across all groups that's just truly amazing," he adds. Yet some listed species end up being more equal than others. Lamoreux explains, for example, that "even if you list a whole lot of dragonflies on the Red List, they're not going to suddenly get as much attention as a panda." He clarifies, accordingly, that "they're not all equal in the eyes of conservation funding or conservation action" (interview). James Watson is president-elect of the Society of Conservation Biology and head of Climate Change Project at the World Conservation Society. Watson points out that of 1,600 species on the Australian threatened list, only 35 percent receive government funding for conservation. "The things which get money are birds and mammals, and the things which don't get money are butterflies and plants," he tells me in an interview. Even the listing of a species as threatened, then, does not promise it equal protection in relation to other listed species. Other criteria, and less formal lists, in fact determine which species are more or less worth saving. Conservation biologist Arne Mooers tells me along these lines that "the conservation biology community [itself] mistakenly considers probabilities of extinction as representing worth" (interview). For this reason, certain conservation biologists have been advocating for alternative or additional lists that justify the differentiation project and make it more scientific and transparent.

Threatened species lists are now everywhere. National agencies routinely make choices on resource allocation among species based on these lists, typically allotting more funding to species listed in the highest threat categories (Possingham *et al.* 2002: 503). Nonetheless, Hoffmann tells me that "there are not many studies that investigate, quantitatively, the impact of listings" (interview). He notes two exceptions: in the United States, recent analyses of recovery plans based on Endangered Species Act listings suggest that there is a positive relationship between funding and trends in species status, and a study of threatened bird recovery programs in Australia for the period between 1993 and 2000 found that where funds have been dedicated to the conservation management of threatened bird taxa, they have produced positive results. "Although more threatened birds declined than increased," the Australian study noted, "many stayed stable over the study period when they might otherwise have become more threatened or gone extinct" (Garnett *et al.* 2003: 664).

The IUCN Red List for Threatened Species™

The Red List is a map of how to do conservation.

(John Lamoreux, interview)

IUCN's Red List is the first modern comprehensive global attempt at listing threatened species. The IUCN has been producing Red Data Books and Red Lists since 1963 (Lamoreux *et al.* 2003: 215). Despite the insistence on the part of many IUCN scientists that the Red List is not prescriptive (Hoffmann, interview), all agree that it has had a profound influence on conservation practices and practitioners around the world (Possingham *et al.* 2002; Rodrigues *et al.* 2006). Specifically, the Red List has inspired the development of numerous national and regional red lists and functions as an important source for the Convention on International Trade in Endangered Species of Wild Fauna and Flora (CITES) – a powerful international convention on trade (Miller 2013) that determines whether and how trade in certain species will be regulated.

The Red List is by far the most influential and widely used method for evaluating global extinction risk. It has been in use for five decades, and has evolved during this period from a subjective expert-based system lacking standardized criteria to a uniform rule-based system (Miller 2013: 195; Mace *et al.* 2008). The IUCN revised its risk-ranking system into data-driven quantitative criteria in 1994 and finalized these categories and criteria in 2001 (IUCN 2001a; see also Mace *et al.* 2008). The current system is designed to provide "a standardized, consistent, and transparent method for assessing extinction risk, thereby increasing the objectivity and scientific credibility of the assessments" (Miller 2013: 195).

The Red List classifies taxa into eight categories: Extinct, Extinct in the Wild, Critically Endangered, Endangered, Vulnerable, Lower Risk, Data Deficient, and Not Evaluated (IUCN 1994). The system consists of one set of criteria that are applicable to all species and that measure the symptoms of endangerment (but not the causes). The three IUCN Red List threatened categories are Critically Endangered, Endangered, and Vulnerable. Five criteria, listed A through E, are used to categorize a taxon within these threatened categories. Although the other categories are formally listed, they are not assessed in the same manner, hence being 'less' listed, or 'list-less.' The threatened criteria are: (A) a reduction in population size; (B) a small, reduced, fragmented, or fluctuating geographic range; (C) a decline in size of an already small population; (D) a very small or restricted population; and (E) a quantitative analysis indicating the probability of extinction. To be listed as Critically Endangered, for example, a species must decline by 90 percent or more, cover less than 100km^2, or consist of fewer than fifty mature individuals (IUCN 2001b). A species need only satisfy one criterion to be listed. Each of these categories contains a list of species, which can be traced in the Red List's online database, with one exception: the category of Not Evaluated includes no taxa (IUCN 2013a), literally establishing a list-less life. List-less, because when a species is not evaluated it is devoid of human protection, thereby remaining mere life. Generally speaking, then, the further the species is ranked away from Extinction, the more unseen it is from the list's perspective and the more killable it is (see Figure 11.1).

Watson says generally about the rigid criteria of the Red List, and of threatened lists more generally, that: "At the end of the day, all listings are arbitrary: they're not driven by the laws of physics, they're actually created ... by humans trying their best to develop the most appropriate categories according to the best available knowledge" (interview). Yet alongside its reliance on fixed and rigid standards, the Red List also enables flexibility and change. Accordingly, the number of species listed in each category changes every time it is updated (on the books, every five years). This is a result of various factors, including species being assessed for the first time, species being reassessed and moved into a different category of threat, and taxonomic revisions. The IUCN distinguishes genuine (namely, real changes in threat levels) from non-genuine (namely, technical changes in threat levels that result from error, taxonomic revisions, or changes in threshold definitions) reasons for revising the listing (IUCN 2013b).

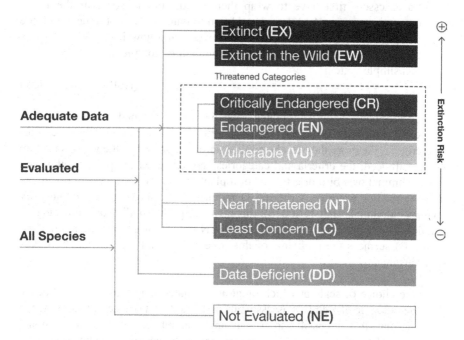

Figure 11.1 The structure of the IUCN Red List Categories (The IUCN Red List of Threatened Species, 2001b © IUCN, reprinted with permission from www.iucnredlist.org/static/categories_criteria_3_1).

The ever-changing nature of the list makes it even more powerful, as no protection, or un-protection, is ever fixed or settled and thus there is constant reliance on the listing process.

In its aspiration to comprehensiveness, simplicity, comparability, consistency, objectivity, and credibility, the Red List is a perfect example of an effective list. By 2013, the IUCN Species Survival Commission network – which is comprised of thousands of scientists and experts from around the world – evaluated the global threat status of 71,576 species of animals, plants, and fungi (IUCN 2013c). The aim: to assess and appropriately categorize every living species (IUCN 2001b). Mike Hoffmann clarifies, accordingly, that the Red List of Threatened Species is in fact not just about threatened species, but about *all* species. "You can't talk about the status of biodiversity globally unless you've assessed everything," he says. Nonetheless, he is first to admit that "we have lots of biases," explaining that the system is "still very much biased towards vertebrates" and that "plants, fungi, and invertebrates are underrepresented" (interview). "We've got a long way to go," he says about the current state of the Red List.

Alongside its comprehensiveness, the Red List is also powerful for its simplicity: "you want a category system that at the end of the day is relatively simple to implement," Hoffmann explains. "There is, already, a fair degree of complexity in the system," he continues:

So assessors first have to wrap their minds around some of the List's common terms.... And then, in addition to that, you've got to deal with the complex biology of your species and understand how it relates to the categories and criteria. So there's a huge amount of complexity already, just in this simple system.

(Hoffmann, interview)

The simplicity factor is intimately related to the heightened comparability that the Red List affords. Generally, the assumption is that the simpler the categories and criteria, the more they can be applied across the board to the various taxa on the list. Indeed, the criteria and categories "are designed to apply whether you are a mammal or a bird or a fungus or a plant or whatever you are" (Hoffmann, interview). For example, Criterion D requires a threshold of fewer than fifty mature individuals (IUCN 2001b); this number applies to all taxa, from fungi to whales. The application of scale in the IUCN criteria of geographic range (Criterion B) surfaces the problems of this 'one size fits all' approach. The IUCN cautions that:

The choice of scale at which range is estimated may thus, itself, influence the outcome of Red List assessments and could be a source of inconsistency and bias. It is impossible to provide any strict but general rules for mapping taxa or habitats; the most appropriate scale will depend on the taxon in question, and the origin and comprehensiveness of the distribution data.

(IUCN 2001b)

Nonetheless, the central idea of the Red List "was to come up with one system that is applicable across all taxa, and you can therefore make *comparisons* across your different taxonomic groups" (Hoffmann, interview, italics added). In addition to the heightened comparability between different taxa, the Red List provides comparability within a particular taxon over time. It makes possible grand calculations such as this one: "On average, 52 species of mammals, birds, and amphibians move one category closer to extinction each year" (see Figure 11.2); or this:

the deterioration for amphibians was equivalent to 662 amphibian species each moving one Red List category closer to extinction over the assessment period, the deteriorations for birds and mammals equate to 223 and 156 species, respectively, deteriorating at least one category.

(Hoffmann *et al.* 2010: 1507)

The Red List's power lies also in its touted objectivity, transparency, and repeatability (namely, that if another expert were to conduct the assessment he or she would reach the same listing conclusion; Brooks, interview). According to Hoffmann, the biggest source of bias is when scientists want to list 'their' species as threatened, "because they're worried that if it's not, they're not going

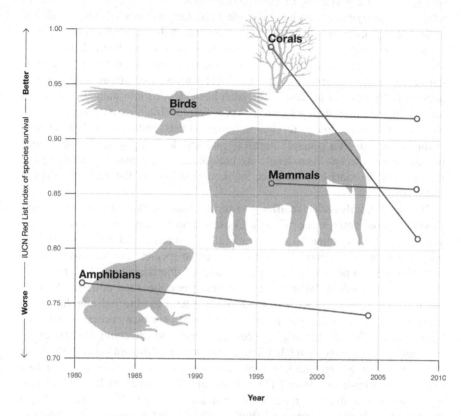

Figure 11.2 The IUCN states that, "Coral species are moving towards increased extinction risk most rapidly, while amphibians are, on average, the most threatened group" (The IUCN Red List of Threatened Species, IUCN 2013b © IUCN, www.iucnredlist.org/about/summary-statistics) reprinted with permission.

to get money." The reverse also happens, with researchers who prefer that their species be listed as Least Concern "so that they can collect their species, put it in a specimen jar, and do research on it." "Our job," Hoffmann tells me, "is to be the neutral, objective, adjudicators of that process." IUCN's Standards and Petitions Subcommittee is the particular adjudicator in cases of disagreement over a Red List designation. According to Hoffmann, they are "the experts in the criteria, and what they say ... would essentially be considered gospel" (interview).

This brings me to the issue of the Red List's credibility. Barney Long is director of Species Protection and Asian Species Conservation at the World Wildlife Fund and a member of the IUCN World Commission on Protected Areas. Long tells me that "when you say this species is red-listed by the IUCN as Critically Endangered, everyone automatically agrees and accepts that. There's no conversation, because the experts have agreed that it is Critically Endangered" (interview). These lists are so important, Long continues, because they are a means for conservationists to communicate with the public, and a source of advice for policymakers on how to protect and manage species. The credibility of the list, it is inferred, creates a front behind which the increasingly fragmented conservation groups can unite, again serving as both a credibility device and a rallying cry. Today, the IUCN Red List is considered one of the most authoritative sources of information on the global conservation status of plants and animals (Lamoreux *et al.* 2003). Its reach has extended into numerous national and international regulatory systems. According to Miller, seventy-six countries use the IUCN methodology for their national red lists (Miller 2013: 197). Hence, "From its origins as a general interest in rare and declining wildlife, the science of threatened species assessment has blossomed into a massive conservation theme with far-reaching influence on conservation on the ground" (Miller 2013: 200).

But there are also adverse affects to certain listings. Brian Horne, turtle conservation coordinator at the Wildlife Conservation Society, tells me in an interview that collectors often "want the rare, and the unusual and different." Hence, when turtle breeders learned that a certain turtle species was soon to be listed under CITES' Appendix I, their prices increased dramatically. "The turtle went from being a hundred dollar turtle to [costing] one thousand dollars." Another result is that once a species is down-listed (the term used to indicate that it has become less threatened), "you become a victim of your own success … because suddenly there's less funding sources available," which could in turn easily translate into less protection (Bennett, interview). In a different context, the price of rhino horn on Korean markets increased by more than 400 percent within two years of their up-listing from CITES Appendix II to Appendix I, which in turn coincided with a sharp increase in the poaching of black rhinos and in illegal trade in rhino horn (Rivalan *et al.* 2007: 530). The listing process thus makes a difference for the lives of animals in myriad, at times counterintuitive, ways.

Listing extinction

On the far end of the Red List continuum, and of threatened species lists more generally, lies the always imminent and looming category of extinction: zero life for the species – as extinction does not apply to individuals. Threatened species lists derive their meaning from this terminal category; it defines both conservation's goal of preventing the extinction of species and its orientation toward crisis intervention. "Extinction [is] the middle name of conservation biology," Redford and his colleagues (2012: 39) write. Furthermore, the category of extinction dictates the lists' focus on rare and threatened species, what Michael Soulé and colleagues (2003: 1247) refer to as "manifest demographic or numerical minimalism." Redford *et al.* (2012: 40) explain that, "[t]his trend is still evident in the fact that successful conservation is defined by many conservation biologists with reference to minimum population sizes, minimum areas, and minimally sufficient sets of sites," which they believe are highly problematic as exclusive measures. This trend emphasizes the high level of trust in numbers on the part of conservationists, a phenomenon that has been problematized more generally by critical population geographers (Legg 2005: 143) and others (e.g., Porter 1996).

But alongside its popular meaning, 'extinction' is also a regulatory term. Indeed, it means one thing in lay discourses, and something quite different in the context of the IUCN Red List. Mike Parr explains that the IUCN has become extremely cautious about listing species as Extinct. The IUCN defines a taxon as Extinct:

> when there is no reasonable doubt that the last individual has died. A taxon is presumed Extinct when exhaustive surveys in known and/or expected habitat, at appropriate times ... throughout its historic range have failed to record an individual. Surveys should be over a time frame appropriate to the taxon's life cycle and life form.
>
> (IUCN 2001b)

Because of these extensive requirements, the last time that a bird was declared extinct was nine years ago (Parr, interview).

Conclusion: listed (and list-less) life

> Both letting die, and making live, have a politics, but I reject the idea that the two are in some kind of functional equilibrium—that it is necessary to select some to die, in order for others to live.... [W]e cannot concede that selection is necessary. It is possible for social forces to mobilize in a wholly make live direction.
>
> (Tania Li 2009: 67)

The last two decades have witnessed an explosion of national lists of threatened and endangered species (see e.g., de Grammont and Cuarón 2006: 22). In 2010, at least 109 countries had produced a national red data book, national red list, or other national list of threatened species (Miller 2013: 198), and at least twenty-five listing systems of threatened species were used across North America (Miller 2013: 192). Of the myriad threatened species lists, Miller writes, some "are designed purely to evaluate risk of extinction, whereas others focus on ranking species to receive priority conservation attention" (Miller 2013: 194).

If the Red List focuses on identifying threatened species, other lists supplement this by identifying alternative targets for maintaining biodiversity. The Alliance for Zero Extinction (AZE) has identified 588 sites that serve as the single remaining location for species listed as Endangered or Critically Endangered under the IUCN Red List (AZE 2013). The AZE is thus an attempt to re-territorialize global threats. Another listing initiative that has emerged in recent years is EDGE of the Zoological Society of London (ZSL), which focuses on red-listed species that possess a significant amount of unique evolutionary history.[1] From the ZSL website: "We have scored the world's mammals and amphibians according to how Evolutionarily Distinct and Globally Endangered (EDGE) they are." "These are the world's most extraordinary threatened species," the website notes, "yet most are unfamiliar and not currently receiving conservation attention" (EDGE 2013).

Alongside the proliferation of lists, a critique of existing listing processes has also emerged. In the words of James Watson: "The conservation field is dominated by ecologists who really like to make lists." But "conservation is also not just about *listing* something," he continues, "it is about *doing* something." "This is not a failure of the list itself," he explains, "it's the failure of the conservation community to develop other metrics beyond the list" (interview, italics added). Joseph *et al.* argue along these lines that existing approaches in conservation typically "ignore two crucial factors: the cost of management and the likelihood that the management will succeed" (2009: 328; see also Bottrill *et al.* 2011; Possingham *et al.* 2002; Walsh *et al.* 2012).

In this chapter, I have explored but a few of the myriad threatened species lists that are currently proliferating in various organizational and regulatory platforms. In particular, I have focused on the IUCN Red List of Threatened Species, the foundation of all modern threatened species lists. Despite their common origin, the various lists differ in their perspective on what is most important about life – and thus what is most worth saving, whether rarity in numbers, unique territorial configurations, or evolutionary (phylogenetic) variation.

Traditionally, animals and plants – along with all that is considered natural or wild – have been confined to the realm of biological life: namely, that which is killable. Conversely, humans have been privileged with political life. This chapter has described how species lists elevate listed nonhuman species from the realm of mere, or biological, life into that of a political life worth saving: laws are put in place to protect life forms belonging to threatened species from being killed or harmed, databases are configured around their most recent census, and those last individuals of such species who die despite the efforts are deeply grieved. Individual life is thereby interpreted and calculated through its configuration and evaluation as a species.

Conservationists believe that life – embodied in species units – must be assessed, listed, and ranked if it is to be protected and saved. The focus of the listing project is thus on life rather than death; this is an affirmative biopolitics that promotes nonhuman survival based on human care and founded on detailed calculations. Conservation is about saving life, but it is also about figuring out which life should be privileged in this endeavor. It begins with the assumption that it is life on the level of the species that should be saved, thereby leaving other life forms to contend with less livable conditions. Such a species life that is worth conserving obtains meaning through an ever-expanding calculus.

Rather than a bifurcated understanding of life versus death, then, conservation lists parse the life of species into much more complicated orderings according to their extinction risks. In other words, the focus of conservationists on authenticity, rarity, and endangerment not only oscillates between life and death, or between political and mere life, but also elevates certain nonhuman species over others, effectively establishing a gradation of animal bodies that are both worth living and worth grieving. Finally, even among those species who are deemed threatened, categories and criteria prioritize the ones who are perceived to be *the most threatened of all*: those whose lives are even more, and finally most, worth saving.

Listing a species gives it a name, a number, a map, and a list of threats – all establishing its uniqueness and elevating it from the list-less abstraction of eco-systems and even populations. The list is the technology through which the species is individuated. My study of threatened species lists thus provides a novel perspective on biopower that highlights its affirmative properties, while at the same time offering a path for critical animal geographers to critically examine the species as a unit of governance in ways that challenge its assumed ontological ordering. Thus we begin to reveal the important messiness of the divide between individuals, species, and populations.

Acknowledgments

I would like to thank Mike Hoffmann, Mike Parr, and Tom Brooks for their time and patience, and Eleanor Gold for her transcription of the interviews and her invaluable editorial help. Research for this chapter was funded by the American Council of Learned Societies' Charles A. Ryskamp Research Fellowship and by the Baldy Center for Law & Social Policy.

Interviews

Bennett, Elizabeth. Vice President for Species Conservation, Wildlife Conservation Society. On-site, New York City, December 20, 2013.

Brooks, Tom. Head, Science and Knowledge Unit, IUCN. Skype, January 25, 2014.

Hoffmann, Michael. Senior Scientific Officer. Species Survival Commission, IUCN. Skype, January 9, 2014; email communication, March 27, 2014.

Horne, Brian. Turtle Conservation Coordinator, Wildlife Conservation Society. Skype, January 9, 2014.

Lamoreux, John. Biodiversity Analyst, National Fish and Wildlife Foundation. Telephone, January 7, 2014.

Long, Barney. Director, Species Protection and Asian Species Conservation, World Wildlife Fund. Skype, January 9, 2014.

Mooers, Arne. Professor of Conservation Biology, Simon Fraser University. Skype, January 6, 2014.

Parr, Mike. Secretary, Alliance for Zero Extinction; Vice President of Planning and Program Development, American Bird Conservancy. Skype, December 23, 2013.

Watson, James. President-elect, Society of Conservation Biology; Head, Climate Change Program, the Wildlife Conservation Society (WCS). Skype, January 27, 2014.

Note

1 The EDGE idea draws on the phylogenetic diversity (PD) concept (Faith 2013). Biodiversity expert Arne Mooers explains that the PD framework could provide a more dynamic – and thus a better – list than EDGE because of its ability to run multiple scenarios with various sets of groups. Mooers provides the example of the kiwi bird from New Zealand to explain the differences between PD and EDGE listings. There are three kiwi species that "aren't related to anything else on the planet," he says. "But even though as a group, they are fifty million years ... distantly related to everything else, amongst themselves they're surprisingly closely related," he explains. "So if you saved any one of them, and let the other two go extinct ... all [of] the 'kiwiness' would still be there, in that one species" (interview). Mooers tells me that all three species rank highly on the EDGE list, resulting in that "you might be wasting your time trying to conserve all three of them, when really you should conserve only one." The kiwi example clarifies the triage function of lists, which operate under the implicit assumption that humans must save certain species rather than others according to their conservation value. "Like in emergency medicine, triage involves using criteria to assess priority and make life or death decisions, not about human beings but about the futures of entire species" (Biermann and Mansfield 2014: 266).

References

AZE 2013, *Alliance for zero extinction*. Viewed November 2, 2013. www.zeroextinction.org/.

Bear, C 2011, 'Being Angelica? Exploring individual animal geographies,' *Area*, vol. 43 (3), pp. 297–304.

Biermann, C and Mansfield, B 2014, 'Biodiversity, purity, and death: Conservation biology as biopolitics,' *Environment and Planning D: Society and Space*, vol. 32 pp. 257–273.

BirdLife International 2014a, 'Cebu Flowerpecker *Dicaeum quadricolor*'. Viewed March 25, 2014. www.birdlife.org/datazone/speciesfactsheet.php?id=8203.

BirdLife International 2014b, 'Cebu Flowerpecker *Dicaeum quadricolor: Additional information*'. Viewed March 25, 2014. www.birdlife.org/datazone/speciesfactsheet.php?id=8203&m=1.

Bottrill, MC, Wash, JC, Watson, JEM, Joseph, LN, Ortega-Argueta, A, and Possingham, HP 2011, 'Does recovery planning improve the status of threatened species?,' *Biological Conservation*, vol. 144 (5), pp. 1595–1601.

Bowker, GC and Star, SL 1999, *Sorting things out: Classification and its consequences*, MIT Press, Cambridge.

Braverman, I 2014, 'Governing the wild: Databases, algorithms, and population models as biopolitics,' *Surveillance & Society*, vol. 12 (1), pp. 15–37.

Braverman, I 2015, *Wild life: The institution of nature*, Stanford University Press, Stanford.

Collar, NJ 1998, 'Extinction by assumption; or, the Romeo Error on Cebu,' *Oryx*, vol. 32 (4), pp. 239–244.

EDGE 2013, *EDGE: Evolutionarily Distinct and Globally Endangered*. Viewed January 26, 2014. www.edgeofexistence.org/.

Faith, D 2013, 'Biodiversity and evolutionary history: Useful extensions of the PD phylogenetic diversity assessment framework,' *Annals of the New York Academy of Sciences*, vol. 1289, pp. 69–89.

Foucault, M 1970, *The order of things: An archeology of the human sciences*, Vintage Books, New York.

Foucault, M 1990, *The history of sexuality: An introduction, volume 1*, Vintage Books, New York.

Foucault, M 2003, *Society must be defended: Lectures at the college de France, 1975–1976*, Allen Lane, London.

Friese, C 2013, *Cloning wildlife: Zoos, captivity, and the future of endangered animals*, New York University Press, New York.

Garnett, S, Crowley, G, and Balmford, A 2003, 'The costs and effectiveness of funding the conservation of Australian threatened birds,' *BioScience*, vol. 53 (7), pp. 658–665.

de Grammont, PC and Cuarón, AD 2006, 'An evaluation of threatened species categorization systems used on the American continent,' *Conservation Biology*, vol. 20 (1), pp. 14–27.

Haraway, D 2008, *When species meet*, University of Minnesota Press, Minneapolis.

Hinchliffe, S and Bingham, N 2008, 'Securing life: The emerging practices of biosecurity,' *Environment and Planning A*, vol. 40 (7), pp. 1534–1551.

Hoffmann, M, Hilton-Taylor, C, Angulo, A, Böhm, M, Brooks, TM and Butchart, SHM et al. 2010, 'The impact of conservation on the status of the world's vertebrates,' *Science*, vol. 330 (6010), pp. 1503–1509.

International Union for Conservation of Nature (IUCN) 1994, *1994 categories & criteria version 2.3*. Viewed April 18, 2013. www.iucnredlist.org/static/categories_criteria_2_3#categories.

International Union for Conservation of Nature (IUCN) 2001a, *Summary of the five criteria (A-E) used to evaluate if a taxon belongs in an IUCN Red List Threatened Category (Critically Endangered, Endangered or Vulnerable)*. Viewed January 26, 2014. www.iucnredlist.org/documents/2001CatsCrit_Summary_EN.pdf.

International Union for Conservation of Nature (IUCN) 2001b, *IUCN Red List Categories and Criteria Version 3.1*, IUCN Species Survival Commission, IUCN, Gland, Switzerland and Cambridge, United Kingdom.

International Union for Conservation of Nature (IUCN) 2013a, *Table 4a: Red List Category summary for all animal classes and orders*, updated November 21, 2013. Viewed January 26, 2014. http://cmsdocs.s3.amazonaws.com/summarystats/2013_2_RL_Stats_Table4a.pdf.

International Union for Conservation of Nature (IUCN) 2013b, *IUCN Red List Summary Statistics*. Viewed January 26, 2014. www.iucnredlist.org/about/summary-statistics.

International Union for Conservation of Nature (IUCN) 2013c, *IUCN Red List of Threatened Species Version 2013.2*. Viewed May 26, 2014. www.iucnredlist.org/search?page=1432.

Joseph, LN, Maloney, RF, and Possingham, HP 2009, 'Optimal allocation of resources among threatened species: A project prioritization protocol,' *Conservation Biology*, vol. 23, pp. 328–338.

Lamoreux, J, Akçakaya, HR, Bennun, L, Collar, NJ, Boitani, L, Brackett, D, Bräutigam, A, Brooks, TM, da Fonseca, GAB, Mittermeier, RA, Rylands, AB, Gärdenfors, U, Hilton-Taylor, C, Mace, G, Stein, BA, and Stuart, S 2003, 'Value of the IUCN Red List,' *TRENDS in Ecology and Evolution*, vol. 18 (5), pp. 214–215.

Legg, S 2005, 'Foucault's population geographies: Classifications, biopolitics and governmental spaces,' *Population, Space and Place*, vol. 11, pp. 137–156.

Li, TM 2009, 'To make live or let die: Rural dispossession and the protection of surplus populations,' *Antipode*, vol. 41 (S2), pp. 66–93.

Mace, GM, Collar, NJ, Gaston, KJ, Hilton-Taylor, C, Akçakaya, HR, Leader-Williams, N, Milner-Gulland, EJ, and Stuart, SN 2008, 'Quantification of extinction risk: IUCN's system for classifying threatened species,' *Conservation Biology*, vol. 22 (6), pp. 1424–1442.

Miller, RM 2013, 'Threatened species: Classification systems and their applications,' in SA Levin (ed.), *Encyclopedia of Biodiversity*, 2nd Edition, vol. 7, pp. 191–211, Academic Press, Waltham, MA.

Oxford English Dictionary Online 2013, Oxford University Press, Oxford.

Porter, TM 1996, *Trust in numbers: The pursuit of objectivity in science and public life*, Princeton University Press, Princeton.

Possingham, HP, Andelman, SJ, Burgman, MA, Medellín, RA, Master, LL, and Keither, DA 2002, 'Limits to the use of threatened species lists,' *Trends in Ecology & Evolution*, vol. 17 (11), pp. 503–507.

Rabinow, P and Rose, N 2006, 'Biopower today,' *Biosocieties*, vol. 1, pp. 195–217.

Redford, KH, Jensen, DB, and Breheny, JJ 2012, 'Integrating the captive and the wild,' *Science*, vol. 338 (6111), pp. 1157–1158.

Rivalan, P, Delmas, V, Angulo, E, Bull, LS, Hall, RJ, Courchamp, F, Rosser, AM, and Leader-Williams, N 2007, 'Can bans stimulate wildlife trade?' *Nature*, vol. 447 (31), pp. 529–530.

Rodrigues, ASL, Pilgrim, JD, Lamoreux, JF, Hoffmann, M, and Brooks, TM 2006, 'The Value of the IUCN Red List for Conservation,' *Trends in Ecology and Evolution*, vol. 21 (2), pp. 71–76.

Rose, N 2001, 'The Politics of life itself,' *Theory, Culture & Society*, vol. 18 (6), pp. 1–30.

Rutherford, S and Rutherford, P 2013, 'Geography and biopolitics,' *Geography Compass*, vol. 7 (6), pp. 423–434.

Sandler, RL 2012, *The ethics of species: An introduction*, Cambridge University Press, Cambridge.

Shukin, N 2009, *Animal capital: Rendering life in biopolitical times*, University of Minnesota Press, Minneapolis.

Soulé, M, Estes, JA, Berger, J, and Martinez Del Rio, C 2003, 'Ecological effectiveness: Conservation goals for interactive species,' *Conservation Biology*, vol. 17 (5), pp. 1238–1250.

Walsh, JC, Watson, JEM, Bottrill, MC, Joseph, LN, and Possingham, HP 2012, 'Trends and biases in the listing and recovery planning for threatened species: An Australian case study,' *Oryx*, vol. 47 (1), pp. 134–143.

Wilcove, DS 2010, 'Endangered species management: The US experience,' in NS Sodhi and PR Ehrlich (eds), *Conservation Biology For All*, pp. 220–235, Oxford University Press, New York.

Wilson, EO 1992, *The diversity of life*, Belknap Press, Cambridge, MA.

Wolfe, C 2013, *Before the law: Humans and other animals in a biopolitical frame*, University of Chicago Press, Chicago.

12 Doing critical animal geographies

Future directions

Rosemary-Claire Collard and Kathryn Gillespie

As I pace the catwalk above the pen where the cow with barcode #743 is lying on the ground, I feel keenly the ethical ambiguity of my research and, more particularly, the problematic dimensions of my presence at the auction. Sitting in the audience at auctions, I position myself in the first or second row of bleachers so that I am at eye level with many of the animals as they move through the auction pen. I watch cows collapse, shake with fear, trip and fall, and bellow to each other across the auction yard. I watch as animals are struck in the face, beaten with rods and shocked with electric prods. And I watch as hundreds of animals are sold without incident or fanfare, destined for the dairy farm, the breeding farm, the slaughterhouse, and the veal crate. My mind races with thoughts of buying even just one of the animals in order to relocate them to a sanctuary and give them a different life. The day-old calves, in a practical way, would be easy to buy (i.e., many of them cost not much more than $15 and, being the size of a large dog, they could easily fit in the back of my station wagon). Instead, I sit there silently. I watch, witness, and document these moments.

. . .

The stacked cages of adult spider monkeys are rolled out of the Lolli Brothers auction ring. I watch from my perch at the far back of the stands and momentarily close the little black notebook in which I record what is bought and sold and for how much. In front of me, several people in the audience hold baby monkeys and baboons in overalls and dresses drinking pop out of straws. One baboon is having its diaper changed. I've been at the auction for hours, watching first buffalos then hundreds of birds – grey parrots, macaws, parakeets, cockatiels, and on and on – as they are auctioned off, many of the birds self-plucked down to goose-pimpled skin, wings clipped, pacing their tiny cages layered with excrement-covered newspaper. The stress and suffering these animals are experiencing is devastating. But all the while I keep recording and work to keep my external reaction neutral. Already marked by appearance, notebook and accent as an outsider, any sign of sadness would make me even more suspicious in a social space outwardly hostile to animal rights. So I stay in place, pen ready to mark the next price, muscles tensed, looking at the backs of hundreds of heads cascading down to the ring and waiting for one to turn and call me out as a spy.

These reflections on doing research at animal auctions bring up the problematic dimensions of witnessing as a research methodology, a topic particularly important for critical animal geographers whose work is motivated by political and ethical commitments to improving the plight of animals in the world. Witnessing is a particular kind of political act – a form of activism – with a long tradition in human, animal, and environmental justice movements. And yet, bearing witness is also fraught with deeply problematic ethical questions: What is our responsibility to the animals we study, especially if and when our research is focused on documenting their suffering? Are we, at least in part, complicit in their suffering by sitting in the audience, watching, documenting, and doing nothing for the individual animals passing through the ring?

This speaks to a larger problem with the act of witnessing and field research in general – namely, that this work is often done in service of longer-term future change, but frequently does nothing for the individual whose suffering is being witnessed. Related to our commitment to interrogating the human–animal divide and speciesism in our introductory chapter: Does this willingness to bear witness reveal something about our ability to maintain species hierarchies even as we do research that actively tries to dismantle these hierarchies? In other words, if our research subjects were human instead of animal, would we feel a more powerful obligation to intervene on their behalf? And are we, as scholars pursuing careers in the academy, using this suffering as a kind of currency in our own career advancement?

This subject of witnessing and its attendant ambiguities also relates to other kinds of ethical, political, and practical questions about doing critical animal geographies. How do we come to know the inner lives of others, particularly when those others are nonhuman animals? How do we gain access to places that are key to animals' experiences of the world, particularly when political and legal mechanisms are actively working to bar access to certain kinds of spaces (see Rasmussen's chapter, this volume, for example)? How are the institutions by which we are employed involved in shaping our research (through funding, human subjects review boards, animal ethics review boards, etc.)?

For example, when the two of us set out to do our most recent individual research projects on cows in the dairy industry (KG) and exotic pets (RC), the course of our fieldwork was impacted profoundly by the way we each responded to certain questions on the ethics review forms. When asked if our research involved animal subjects, one of us (RC) replied "no" and the other (KG) replied "yes." Checking the 'yes' box led to a long and involved process of gaining approval from the Institutional Animal Care and Use Committee (IACUC) for research involving animals. And yet, the process of ethics approval was designed explicitly (and fairly exclusively) for *biomedical* research involving animals. As such, part of the approval process involved completing an online training course on working with animals in research, focused almost entirely on approved forms of euthanasia for animals at the end of the study. So while there was ample information on decapitation as the approved method of killing rats in a laboratory, there were no ethical guidelines for encountering cows or exotic pets at the auction yard in social science fieldwork. As we have argued elsewhere, this example speaks to a larger problem in the academy, which is that academic research on animals – the kind that is focused on advancing (not mitigating) their autonomy – is not legible in current institutional structures of ethical oversight and research design (Gillespie 2014). Current forms of ethical review are exemplars of a broader institutional orientation that settles around the human–animal divide. Animals are considered outside the purview of 'human' ethics, and animal ethics revolves, in most cases, around a presumed 'disposable' animal life (Collard 2015).

That much animal research occurs within a sensitive and ethically and politically charged context compounds this problem. As some scholars have described, carrying out research in a "hostile environment" or "behind enemy lines" can be frustrating, frightening, anxiety ridden, and exhausting (Han 2010; Gould 2010). Animal researchers in particular have few supports – institutional, emotional – for dealing with ethical dilemmas that can therefore arise in the field and through the experience of witnessing (or participating in) animal suffering. The enduring anthropocentrism in academia and beyond can lead even those researchers who argue for the recognition of animal subjectivity to deny or question the ethical and emotional effects of conducting research on and with animals. As Stănescu (2012: 2) writes, to care "for the existence of beings whom most people manage to ignore ... to tear up, or to have trouble functioning ... is something rendered completely socially unintelligible [and so] most of us work hard not to mourn." In particular field sites, like the public animal auction houses we describe above and in the introductory chapter, it can feel that grief over animal suffering is "out-of-place" (Gillespie 2014). Expressing such grief would be a marker of one not belonging in these spaces – spaces host, for example, to hostile relations between animal owners and animal activists, as animal auctions are. Working not to mourn can add further stress to the already difficult practice of repeatedly observing animal suffering and death.

Future directions for critical animal geographies

Above we have described some of the challenges of conducting primary research in what can be thought of as "multispecies contact zones," following Mary Louise Pratt (1991, 1992; also Haraway 2003, 2008). For Pratt (1991) contact zones are where two or more cultures intermingle; they are "social spaces where cultures meet, clash, and grapple with each other, often in contexts of highly asymmetrical relations of power" (Pratt 1991: 34). In *Imperial eyes*, Pratt (1992: 7) writes that as a thinking device and methodological frame, the contact zone

> emphasizes how subjects are constituted in and by their relations to each other. It treats the relations among colonizers and colonized, or travelers and 'travelees,' not in terms of separateness or apartheid, but in terms of co-presence, interaction, interlocking understandings and practices.

Pratt is focused on asymmetries among humans, but her work finds echoes in Donna Haraway's (2003, 2008), which is resolutely multispecies. For Haraway, acknowledgement of "the foolishness of human exceptionalism" – acceptance of our inextricable entanglement with human and nonhuman others – means "being" human is always a "becoming-with" a multitude of others. This occurs within contact zones that for Haraway, like Pratt, are always saturated with power.

Research, too, occurs in these multispecies contact zones. Indeed as researchers we often seek out spaces in which we will have direct sensory engagement with animals, where animals become partners in our research practice. But the animals we encounter in contact zones are usually "significantly unfree partners" (Haraway 2008). Much of the time they are captive, or at the very least subject to varying degrees of bodily or spatial enclosure and manipulation. They have no choice but to be encounterable and researchable. Animal research subjects do not and cannot give consent; in fact they may demonstrate an active desire not to participate in research. This desire is usually overridden. It is hard to imagine a more pronounced power asymmetry between researcher and research subject. Researchers may therefore experience significant internal unrest about not only witnessing but also at times actively interacting with animals who are clearly suffering, some to the point of death, before their eyes and in their hands.

Moreover, when studying animals (and their encounters) with participant- or spectator-observation in multispecies contact zones, we are actors within those zones, and so our research practices often bring us into direct complicity with the power dynamics we are seeking to contest. As Timothy Pachirat (2011: 16) notes in his undercover ethnography of a Nebraska slaughterhouse, "once inside as an active participant I found myself inextricably caught up in its networks of power." Although Pachirat is writing of the slaughterhouse's highly racialized, classed, and gendered labor distribution, he expresses similar complicity with the slaughter itself, even noting at the outset of the book, in his acknowledgements, that he aided in taking at least 240,000 animals' lives during his half a year laboring at the slaughterhouse. This complicity can engender a great deal of ambivalence.

The root of this ambivalence is an ongoing methodological dilemma. Researchers are not flies on the wall in contact zones. When conducting research we are caught up in networks of power within which we are actors, not passive observers. Research may be marked by a central struggle between seeking out encounters with animals for research purposes, and knowing that conditions of violence, suffering, and asymmetrical power make these encounters possible. This tension remains unexplored in nascent conversations about 'more-than-human methodology' in geography (Whatmore 2006; Lorimer 2010). Given the current enthusiasm for animal research, we hope that more sustained conversation develops about methodological considerations.

In doing so, critical animal geographers could learn much from engaging multispecies ethnography as a methodology in their research. Multispecies ethnographies are a growing trend in anthropology where "ethnographers are studying the host of organisms whose lives and deaths are linked to human social worlds [and whose lives] shape and are shaped by political, economic and cultural forces" (Kirksey and Helmreich 2010: 545). This methodology invites inquiry into the intimately intertwined enactments of violence, power, cruelty, love, care, and kinship that define the varied and ambivalent relationships between humans and other animals. That it is multispecies allows for critical analysis of not only animals and their worlds, but also the complex interactions occurring among humans and other species. Critical animal geographers offer unique contributions to this burgeoning field of scholarship. That is, they are attentive to the places and spaces (or rather, the multispecies contact zones) in which these cross-species encounters occur. Attention to these contact zones and to the fraught power relations existing in them is a key feature in a critical geographical multispecies ethnographic approach.

We hope that, in the coming years, critical animal geographers will take up the task of exploring the limits and possibilities of current multispecies methodologies, as well as innovating new methodologies that attend to these ethical and political ambiguities. We also see critical animal geography as poised to contribute to several other exciting lines of inquiry in the broader conversation about more-than-human ways of being in the world. In a spirit of encouragement and enthusiasm, we dedicate the remainder of the chapter to offering our ideas about some of these possible paths forward.

Teaching critical animal geographies

As the field of critical animal geographies (and critical animal studies more generally) grows, universities and colleges are already becoming more open to including courses on animals (or with an animal component) in their curricula. Even in our teaching experience in the past five or so years, students have shown great interest in learning about animals and engaging with complex ethico-political questions about animals' experience of the world and humans' responsibility to them. In response, new writing, such as Margo DeMello's (2010) anthology *Teaching the Animal*, focuses on how to teach effectively about animals and human–animal relations across the disciplines. In this vein, how can geography take up dynamic and effective modes of teaching critical animal geographies? What can a geographical approach to pedagogy add to the practice of teaching and learning in a multispecies world?

To illustrate one example of what an unconventional way of 'teaching the animal' might look like, one of us (KG) taught an interdisciplinary summer course in 2014, *Animals, Ethics and Food: Doing Multispecies Ethnography*, which spent one full day a week at a local sanctuary for pigs. In addition to reading, viewing documentaries, and engaging in discussion in the classroom, the students were each paired with a pig for the quarter and conducted mini-ethnographies at the sanctuary. Integral to their analyses was a keen attention to the geography of the sanctuary – in other words, coming to understand the sanctuary *as a place* and the importance of that place for the human–animal encounters that were enacted there. As a core feature of this course, the act of 'being with' – of coming face to face with – animal others at the sanctuary did more in just a few short weeks than a full term of reading and theorizing alone could have accomplished toward encouraging students to think about species difference and uneven structures of power. As critical animal geographers have the opportunity to teach about animals more frequently in the coming years, discussions and collaborations with other critical animal geographers will be increasingly important in order to imagine new, creative approaches to pedagogy that are attentive to different ways of being in a multispecies world.

Non-universalizing approaches

A recent intervention by Juanita Sundberg (2013) emphasizes that in geographers' enthusiasm to decenter the human, there is a risk of universalizing their subject of critique – that is, a particular figure of the human that is bounded, rational, superior, and transcendent. Sundberg argues that this figure of the human is a colonial figure, one particular to Eurocentric modes of thought. Kay Anderson (2007, 2014) carefully fleshes out this point by tracking the production of the colonial human subject through not only discursive but also material practices, such as craniometry. As she argues, race is a key marker or classification deployed within efforts to mark out this figure of the human, particularly following the gradual decline of the divine order, the Great Chain of Being. Failing to recognize the coloniality, racial contours, and specificity of this human figure can lead to critiques that are complicit in wider disavowals or subordinations of other ways of thinking and being. Future work in critical animal geography can learn from scholars in geography and other disciplines who take this critique seriously and work carefully toward specificity and non-universalizing approaches (e.g., Rose 2004; Anderson 2007; Ahuja 2009; Huggan and Tiffin 2010; Gaard 2013; Garcia 2013).

One recent effort dedicated to the difficult task of negotiating conflicting human and animal interests, particularly in emotionally charged cultural contexts, is Claire Kim's *Dangerous Crossings*. Kim's (2015) work explores three sites of multispecies cultural conflict (e.g., animal rights activists versus Makah community members interested in a return to whaling; animal rights activists versus Chinese live animal market vendors and consumers; and conflicting racialized responses to Michael Vick's dog-fighting ring). Racist tropes often accompany critiques of cultural practices involving animals; for this and other reasons, these conflicts are usually characterized by a process of "mutual disavowal" for the competing positions. Buried within these human cultural conflicts, animals often lose. As a path forward, Kim (2015) urges us to move toward a position of "mutual avowal" where all conflicting human positions are recognized and internalized, and where the animal subject also has a significant interest represented in the conversation. Thus, critical animal geographies might make inroads to interrogate these difficult sites of negotiation – these "dangerous crossings" as Kim calls them – in order to advance geographical research trajectories which negotiate and contest these universalizing trends while still centering the animal in our studies.

Potential topics for future research

Finally, as critical animal geographies expands as a field, there are many exciting new directions in empirical research that critical animal geographers might explore.

One thread of research which we hope critical animal geographers might undertake more centrally is the varied and ambivalent use of animals in our everyday lives. In addition to growing attention on animals raised for food (Emel and Neo 2011; Gillespie 2013; Blecha and Leitner 2014), critical animal geographers might turn their gaze to the subject of animals used in garment production and the fashion industry more generally. The fur, leather, wool, feather, and silk industries are vastly understudied in and beyond geography, particularly with regards to the impacts of these industries on animals. Wearing animals is, like eating animals, an everyday enactment of violence against animal bodies – an act of dominance – and is made invisible by the extent to which these 'materials' are ubiquitous in our lives.

In this same vein of everyday and invisibilized use of animals, work that critically examines the role of animals in medical research and product testing is urgently needed. Dogs, cats, nonhuman primates, pigs, rats, mice, guinea pigs, rabbits, ferrets, frogs, and other amphibians continue to languish in research laboratories in the name of science, bearing not only "the burden of our dreams of health and longevity" but also assuming "the task of living out our nightmares," as Hugh Raffles (2010: 120) says. Thus, critical animal geographers should take up the project of interrogating the power geometries in the research lab with careful attention to the different ways in which violence is enacted on animals in service to human medical advancements. As this is a highly contentious and politically charged topic, especially at institutions which receive significant funding for biomedical research, it will be particularly important for more senior scholars, especially those with tenure, to undertake this research.

Although animal geographers have a strong tradition of research in human–wildlife relations, some key institutions and practices in terms of wildlife 'management' have gone under-explored. Wildlife rehabilitation is an increasingly popular interface between humans and wild animals, yet it has received little critical scrutiny. Several assumptions tend to be operationalized in wildlife rehabilitation, prime among them that a captive life is preferable to a death outside of cage walls, and that the animals' 'natural' disposition is to fear humans (Collard 2014). These biopolitical and misanthropic assumptions deserve greater debate. Additionally, when animals are released from rehabilitation, they are often equipped with tracking and monitoring devices, a common scientific practice more broadly. Tracking technologies and their effects are another understudied topic in animal geography. Key questions remain to be explored: does the concept of privacy have any purchase or place in thinking about tracking, or recording animals more generally, for example on film? What is produced through modes of animal surveillance? Do they reconfigure human–animal relations, and if so, how? What are the power dynamics of these knowledge generation projects?

Finally, some promising moves are being made in animal geography to broaden the scope of inquiry to include less charismatic or "awkward creatures" (Ginn *et al.* 2014; Ginn 2013). What are our responsibilities to "unloved others" (Rose and van Dooren 2011), the scavengers, rodents, insects, and other animals deemed 'pests,' animals who are born exterminatable and who are subject to everyday forms of violence? These are animals who are deeply interwoven in human socio-ecologies and economies, but whose roles are frequently overlooked. These are also animals whose complex social worlds tend to go less recognized than charismatic mammals, although as Raffles's (2010) intricate, playful, and uncanny *Insectopedia* shows us, this is in large part a failure of imagination on our parts.

Conclusion

Our hope for critical animal geographies in the coming decades is that the field will continue to grow and interrogate the violent and power-laden relationships between humans and other animals. We urge critical animal geographers to engage with other fields dedicated to issues of social, environmental, and animal justice in order to collaborate on broader issues of justice and deconstruct hierarchical social relations in various contexts. Our ambivalence about the ethics of doing research with and for animals in fieldwork should not dissuade others considering such work. As feminist geographer Liz Bondi (2004) and others (Haraway 1991; Rose 1991; McDowell 1992) have written, ambivalence can be a key political strategy. For Bondi (2004: 5), the position of the feminist academic "is a contradiction in terms" given that feminists are required "to take up a position of authority at the same time as acknowledging the fraudulence of claims to such a position." Similarly, to be a researcher who is at once critical of the human–animal divide and dominant human practices toward animals and is an academic is to inhabit a paradoxical space. Higher education institutions have historically been leaders in concertedly humanist thought. To the extent that animals appear in academic landscapes, it is generally as scientific test subjects, confined to tightly secured and clandestine laboratories. Academia is by and large a humanist, speciesist space. This makes the anti-speciesist academic, like the feminist one, a contradiction in terms.

For Bondi, however, this ambivalence is not something to conquer. It should be mobilized in a "politics of ambivalence ... [that is] not about 'sitting on the fence,' but about creating spaces in which tensions, contradictions and paradoxes can be negotiated fruitfully and dynamically" (Bondi 2004: 5). Multispecies contact zones are examples of such spaces. Negotiating tensions within them is an indelible part of the research experience, and can illuminate the existence of multiple worlds. In other words, the paradoxes of research experience – and of being an anti-speciesist academic – can help to disrupt assumptions and intellectual comfort zones. This does not mean researchers should conduct research anywhere they want, regardless of the consequences. It rather points to the need to be relentlessly attentive to the conditions that make our research experiences possible, and to the effects of our "research performances" (Pratt 2000). With this attention to the politics of ambivalence in mind, we hope that future critical animal geographers will embark on research projects that dismantle hierarchy, acknowledge different ways of being in the world, and work with increasing intensity for more just interspecies social relations.

References

Ahuja, N 2009, 'Postcolonial critique in a multispecies world,' *PMLA*, vol. 124, pp. 556–563.

Anderson, K 2007, *Race and the crisis of humanism*, Routledge, New York.

Anderson, K 2014, 'Mind over matter? On decentring the human in Human Geography,' *Cultural Geographies*, vol. 21, pp. 3–18.

Blecha, J and Leitner, H 2014, 'Reimagining the food system, the economy, and urban life: new urban chicken-keepers in US cities,' *Urban geography*, vol. 35, pp. 86–108.

Bondi, L 2004, '10th Anniversary Address: For a feminist geography of ambivalence,' *Gender, Place & Culture*, vol. 11, pp. 3–15.

Collard, R-C 2014, 'Putting animals back together, taking commodities apart,' *Annals of the Association of American Geographers*, vol. 104, pp. 151–165.

Collard, R-C, 2015, 'Ethics in research beyond the human,' in T Perrault, G Bridge, and J McCarthy (eds), *The handbook of political ecology*, Routledge, New York.

DeMello, M 2010, *Teaching the animal: Human–animal studies across the disciplines*, Lantern Books, New York.

Emel, J and Neo, H 2011, 'Killing for profit,' in R Peet, P Robbins, and M Watts (eds), *Global political ecology*, Routledge, New York.

Gaard, G 2013, 'Toward a feminist postcolonial milk studies,' *American Quarterly*, vol. 65, pp. 595–618.

Garcia, ME 2013, 'The taste of conquest: Colonialism, cosmopolitics, and the dark side of Peru's gastronomic boom,' *The Journal of Latin American and Caribbean Anthropology*, vol. 18, pp. 505–524.

Gillespie, K 2013, 'Sexualized violence and the gendered commodification of the animal body in Pacific Northwest US dairy production,' *Gender, Place and Culture*, early view.

Gillespie, K 2014, 'Reproducing dairy: Embodied animals and the institution of animal agriculture,' Doctoral Dissertation, University of Washington, Seattle.

Ginn, F 2013, 'Sticky lives: Slugs, detachment and more-than-human ethics in the garden,' *Transactions of the Institute of British Geographers*, early view.

Ginn, F, Biesel, U, and Barua, M 2014, 'Flourishing with awkward creatures: togetherness, vulnerability, killing,' *Environmental Humanities*, vol. 4, pp. 113–123.

Gould, K 2010, 'Anxiety, epistemology, and policy research "behind enemy lines",' *Geoforum*, vol. 41, pp. 15–18.

Han, J 2010, 'Neither friends nor foes: Thoughts on ethnographic distance,' *Geoforum*, vol. 41, pp. 11–14.

Haraway, D 1991, *Simians, cyborgs, and women: The reinvention of nature*, Routledge, New York.

Haraway, D 2003, *The companion species manifesto: Dogs, people, and significant otherness*, Prickly Paradigm Press, Chicago.

Haraway, D 2008, *When species meet*, University of Minnesota Press, Minneapolis.

Huggan, G and Tiffin, H 2010, *Postcolonial ecocriticism: Literature, animals, environment*, Routledge, London and New York.

Kim, C 2015, *Dangerous crossings: Race, species, and nature in a multicultural age*, Cambridge University Press, New York.

Kirksey, SE and Helmreich, S 2010, 'The emergence of the multispecies ethnography,' *Cultural Anthropology*, vol. 25, pp. 545–576.

Lorimer, J 2010, 'Moving image methodologies for more-than-human geographies,' *Cultural Geographies*, vol. 17, pp. 237–258.

McDowell, L 1992, 'Multiple voices: Speaking from inside and outside "the project",' *Antipode*, vol. 24, pp. 56–72.

Pachirat, T 2011, *Every twelve seconds: Industrialized slaughter and the politics of sight*, Yale University Press, New Haven, CT.

Pratt, G 2000, 'Research performances,' *Environment and Planning D: Society and Space*, vol. 18, pp. 639–651.

Pratt, ML 1991, 'Arts of the contact zone,' *Profession*, vol. 91, pp. 33–40.

Pratt, ML 1992, *Imperial eyes: Travel writing and transculturation*, Routledge, New York.

Raffles, H 2010, *Insectopedia*, Pantheon, New York.

Rose, DB 2004, *Reports from a wild country: Ethics for decolonisation*, UNSW Press, Sydney, NSW.

Rose, G 1991, 'On being ambivalent: Women and feminisms in geography,' in C Philo (ed.), *New words, new worlds: Papers and proceedings*, pp. 156–163, Social and Cultural Geography Study Group of the Institute of British Geographers, St. David's University College, Lempeter, Wales.

Rose, DB and van Dooren, T 2011, 'Unloved others: Death of the disregarded in the time of extinctions,' *Australian Humanities Review*, vol. 50. Viewed September 17, 2014. www.australianhumanitiesreview.org/archive/Issue-May-2011/home.html.

Stănescu, J 2012, 'Species trouble: Judith Butler, mourning, and the precarious lives of animals,' *Hypatia*, vol. 27, pp. 567–582.

Sundberg, J 2013, 'Decolonizing posthumanist geographies,' *Cultural Geographies*, vol. 21, pp. 33–47.

Whatmore, S 2006, 'Materialist returns: Practising cultural geography in and for a more-than-human world,' *Cultural Geographies*, vol. 13, pp. 600–609.

Index

Page numbers in **bold** denote figures.

Printed in the United States
by Baker & Taylor Publisher Services